Najeh TKA

Synthesis and characterization of new chiral aminohydrazides

Najeh TKA

Synthesis and characterization of new chiral aminohydrazides

Natural amino acids are easily converted to aminohydrazides

ScienciaScripts

Imprint
Any brand names and product names mentioned in this book are subject to trademark, brand or patent protection and are trademarks or registered trademarks of their respective holders. The use of brand names, product names, common names, trade names, product descriptions etc. even without a particular marking in this work is in no way to be construed to mean that such names may be regarded as unrestricted in respect of trademark and brand protection legislation and could thus be used by anyone.

Cover image: www.ingimage.com

This book is a translation from the original published under ISBN 978-620-6-70705-9.

Publisher:
Sciencia Scripts
is a trademark of
Dodo Books Indian Ocean Ltd. and OmniScriptum S.R.L publishing group

120 High Road, East Finchley, London, N2 9ED, United Kingdom
Str. Armeneasca 28/1, office 1, Chisinau MD-2012, Republic of Moldova, Europe
Printed at: see last page
ISBN: 978-620-8-35727-6

Contents

General introduction

Chirality is an important concept in many areas of science, being fundamental to the chemistry of living organisms and indispensable for certain biological properties.1

In 1848, Louis Pasteur discovered the concept of chiralite[2] by separating the two forms of tartaric acidlet **1' (Figurel).** Louis Pasteur also demonstrated that asparagine, isolated from vetch plants, is optically active and that the aspartic acid produced by hydrolysis is also optically active.

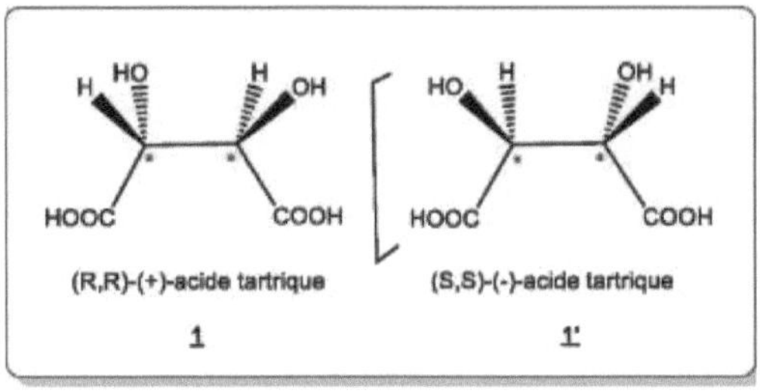

Figure 1

This chirality problem has proved to be of vital importance in the development of modern chemistry and biology. Shortly afterwards, in 1849, Puitti noted that (L)-asparagine2 tastes bitter while its enantiomer (D) tastes sweet[3,4] **(Figure2).**

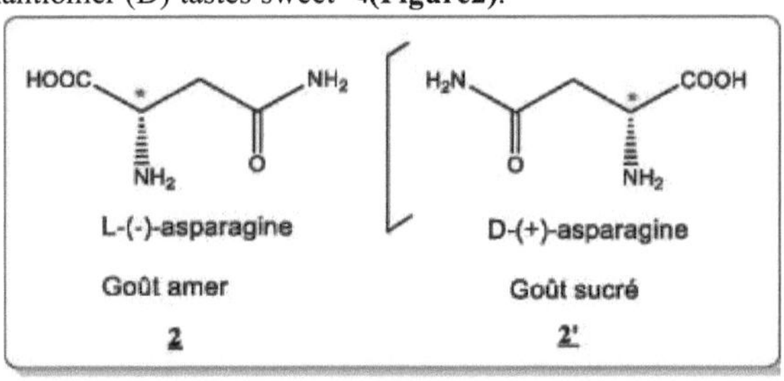

Figure 2

Another example illustrating the different activities of enantiomers is carvone. Carvone exists in nature in the form of two optical antipodes: the R-(-)-**carvone3** isomer, the main component of spearmint essential oil, is responsible for the smell and taste of mint[5] the antinociceptive activity[6] and the antispasmodic effect.7

Whereas S-(+)-**carvone3'**1 ,found in caraway essential oil, produces the characteristic caraway odour[8] **(Figure3).**

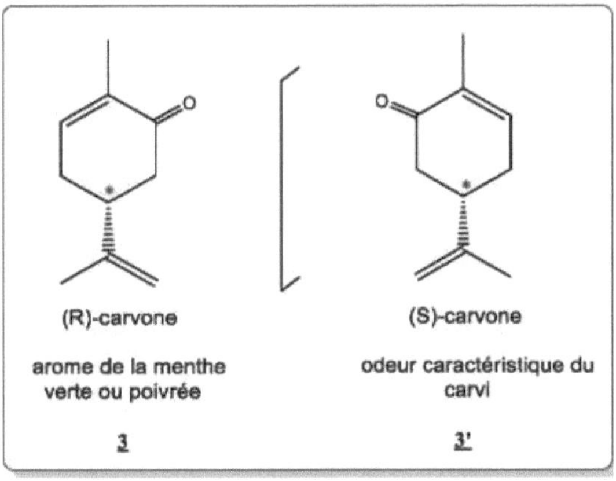

Figure 3

Two possëdent enantiomers have identical physicochemical properties in a symmëtrical environment. However, they are perceived differently by living organisms. In other words, depending on whether the molecule is in one form or another, it will not have the same effect on our body, as is the case for certain drugs. Chiodini et al.[9] noted that (S)-oxiracetam4 can act as a neuroprotectant, whereas its enantiomer (R) is inactive10**(Figure 4).**

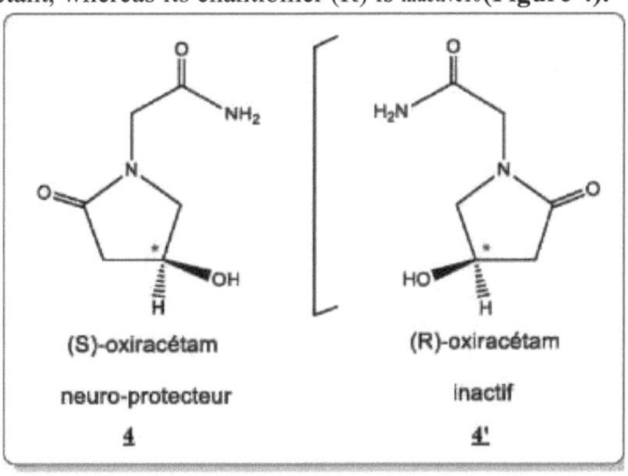

Figure 4

Another example is dextromethorphan5_, a drug used to treat coughs, while its antagonist lelevomethorphan5' is a powerful narcotic[n] **(Figure5).**

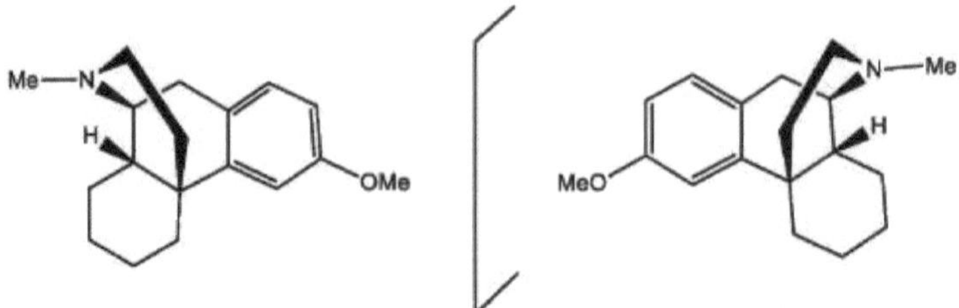

dextromethorphan
médicament contre la toux
5

levomethorphan
narcotique puissant, analgésique
5'

Figure 5

Napropamide, for example, is a herbicide currently sold in the form of a unmëlange racëmique. Recently, work by Yanli Qi et al.[12] has shown that R-(-)-napropamide6 affects plant growth more severely than S-(-)-napropamide.

(+)-**napropamide6' (Figure 6).**

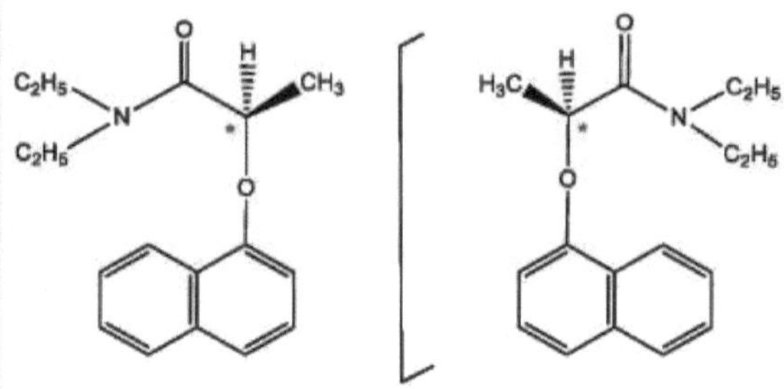

R-(-)-napropamide
6

S-(+)-napropamide
6'

Figure 6

This constraint clearly demonstrates the importance of strict control of the stërëochimy of chemical products. At present, the molecules used in pharmaceuticals are increasingly developed in the form of a single enantiomer. Nowadays, there are three main techniques for obtaining optically pure chiral compounds:

- The use of a chiral pool.
- The doubling of the racemic mix.
- Asymmetric synthesis.

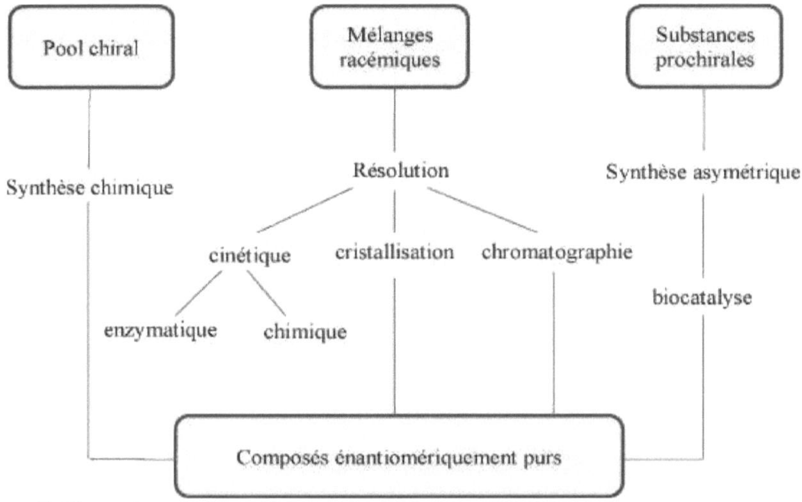

Figure 7: The main techniques for accessing enantiopure molecules

The modification of substrates from the so-called chiral pool certainly remains one of the best methods for accessing chemical compounds with very high enantiomeric purity. These chiral substrates have a major advantage in that their stereochemistries are known and, for the vast majority of them, the enantiomeric purity is often preserved.

The research work carried out in this thesis is linked to this method and aims to synthesise new chiral N-phenylsulphonyl-a-aminohydrazides from natural amino acids. The present work can be divided into three parts:

- The first part is devoted to the synthesis of a series of chiral N-phenylsulphonyl-a-amino acids from naturally occurring a-amino acids.

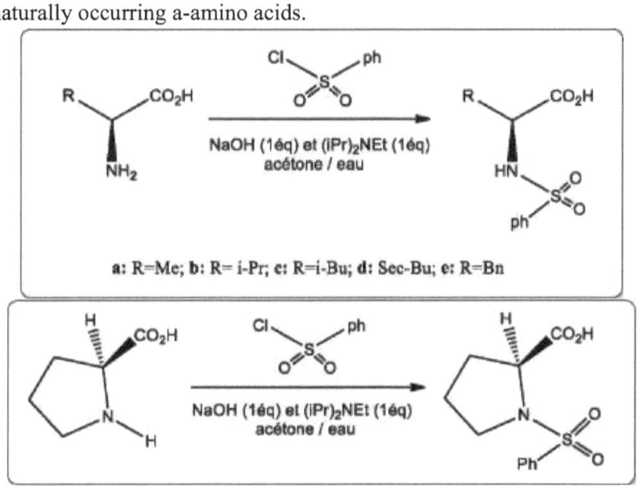

The second part describes the synthesis of chiral N-phenylsulphonyl-a-amino *esters* by esterification of the N-phenylsulphonyl amino acids obtained previously.

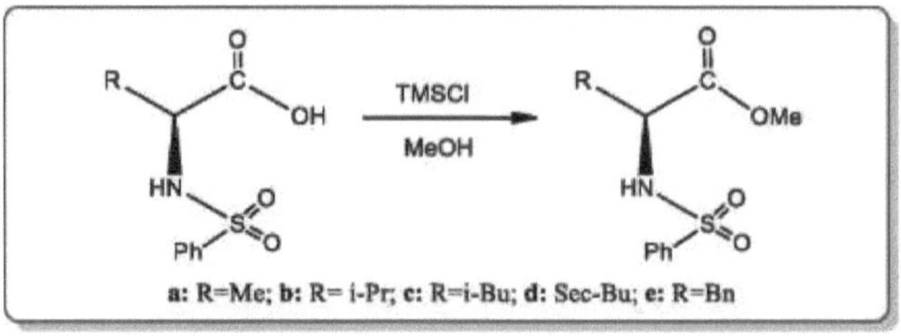

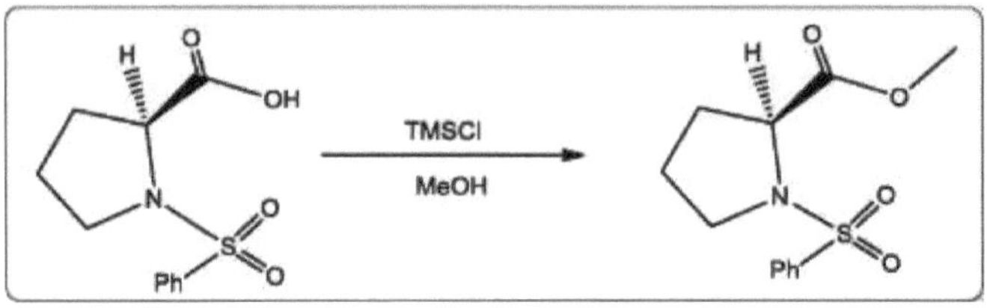

The third part deals with the synthesis of chiral N-phenylsulphonyl-a-aminohydrazides by condensation of the a-amino esters obtained with 1'hydrazine.

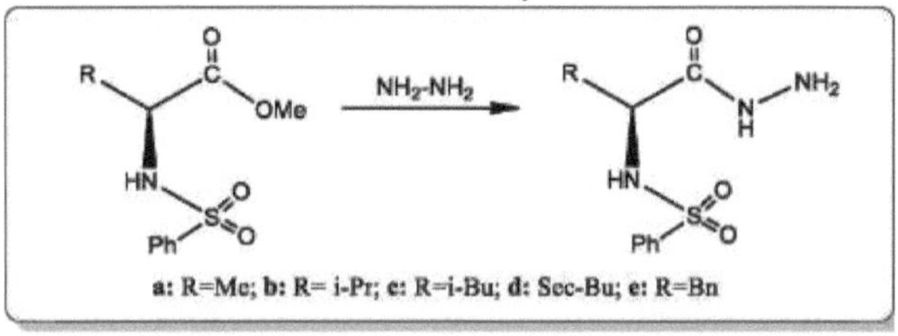

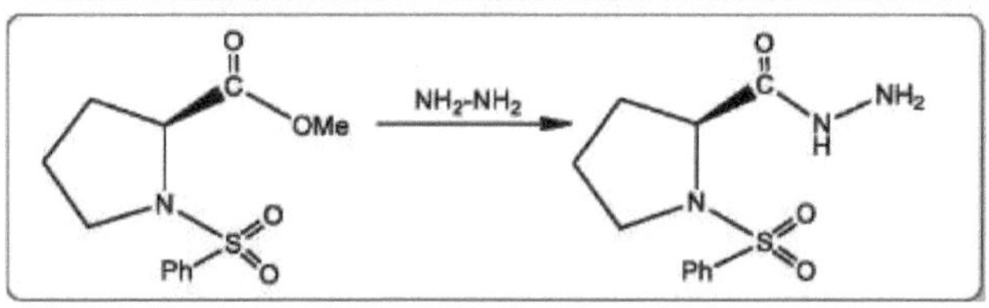

Chapter I

Synthesis of N-phenylsulfonyl-a-amino *acids* from natural amino acids

A. General information and bibliographical references

Amino acids are compounds with a basic amine function and an acidic carboxylic acid function **(Figure 8)**.There are two families of amino acids:

- A-amino acids **7**: the two amine and carboxylic acid functions are carried by the same carbon.
- E-amino **acids8**: the acid function is carried by the carbon adjoining the carbon carrying the amine function.

7
α-aminoacide

8
β-aminoacide

Les groupes R_1,R_2,R_3,R_4 représentent les chaînes latérales des acides aminés

Figure 8

In recent years, the use of amino acids as precursors to access chiral molecules has developed well[13-14] . Natural amino acids9 , which have an absolute configuration of the (L) type in the Fischer representation[15] , are more widely used than their non-natural counterparts (D)10 , **which** are expensive and difficult to access[16] **(Figure 9)**.

(L)-aminoacide
9

(D)-aminoacide
10

Figure 9

Often, amino acids are represented in ionic form because the acid is dëprotonë and on the other hand the amine is protonated**(Figure 10)**. The amine group exerts an attractive inductive effect that leads to a атëlюraьоп of the acidit^ by increasing the polarisation of the (OH) bond, which then becomes more fragile.The intramolecular hydrogëne bond that forms between the oxygen and the hydrogëne stabilises the amino acids considerably.

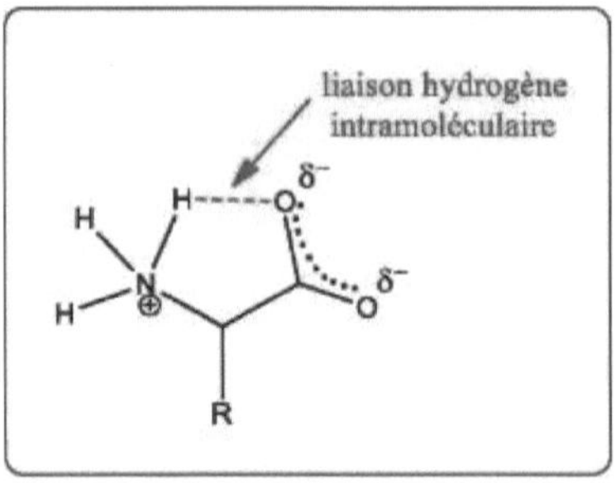

Figure 10

I. Different methods for protecting the amine function of amino acids

To work easily on the COOH carboxyl group of amino acids[17] , the amine function must be protected by a protective agent.

The amine function must be protected by a group that is stable to most chemical manipulations and can be easily removed under mild conditions. The most commonly used reagents for protecting the amine function are carboxylic acid derivatives (chlorides or anhydrides). Each protective group possëde a specific stability, binding and elimination[18] . Some examples are given below:

I.1 Protection by t-butyloxycarbonyl (BOC)

T-butyloxycarbonyl is the most widely used group for protecting amines.[19] The first use of this group with amino acids, during peptide synthesis, was by Mckay et al.[20-21] **(Scheme 1).**

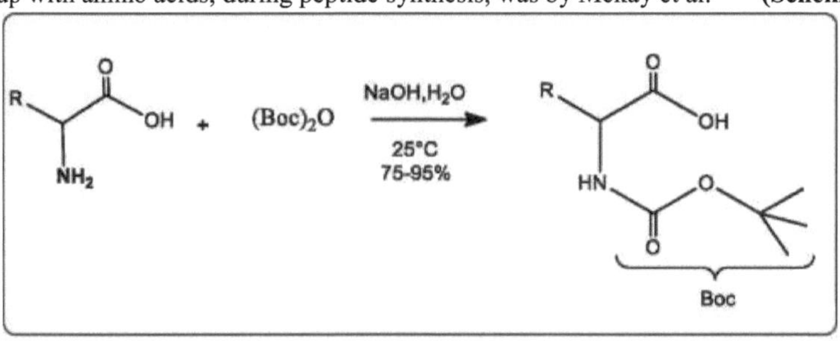

Schéma 1

(Boc)2Ois often етрloyë in an aqueous sodium hydroxide solution. The mëchanism of the protection is represented^ in the following sclu'ma**(Schema 2)**:

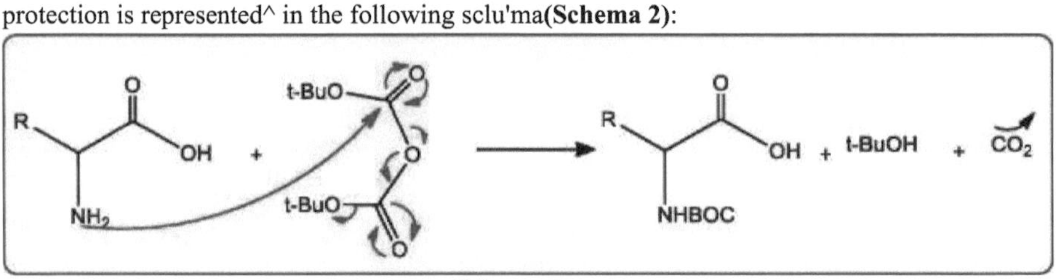

Schema 2

I.2 Protection by 9-fluorenylmethoxycarbonyl (Fmoc)

9-fluorenylmëthoxycarbonyl(Fmoc) is very stable under acidic conditions. It is widely repanded in peptide chemistry and peptide solid phase chemistry. A.Caprino and Y.Han ШШЗё this group for the first time in protecting the amine function. The (Fmoc) group is stable under acidic conditions and resists catalytic hydrogënation. Its cleavage is carried out under mild basic conditions in the presence of a base such as pyridine or diisopropy^thylamine (iPr)2NEt at room tempërature[22-23] **(Schema 3)**.

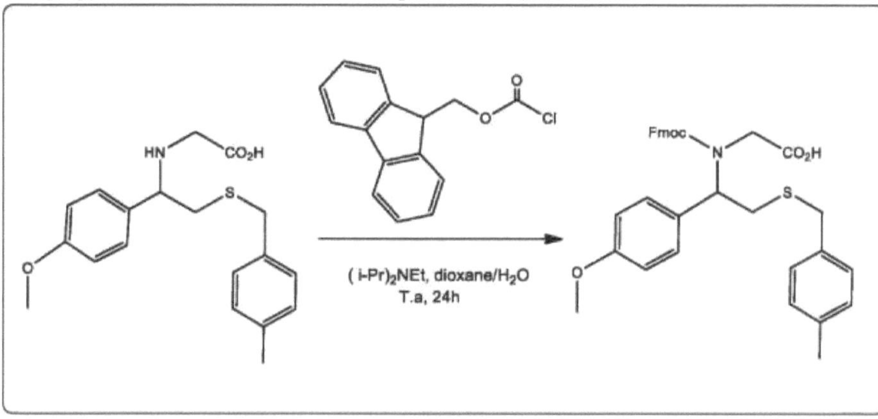

Schéma 3

I.3 Protection by benzyloxycarbonyl (Cbz)

Benzyloxycarbonyl (Cbz) is the most widely used protecting group for the amine function because it is very stable under acidic conditions and can be easily eliminated by several methods. *N-benzyloxycarbonyl-a-amino* **acids13** are obtained by the action of benzyl **chloroformate12** on the corresponding **amino** acidsll[24-25] **(Schema 4).**

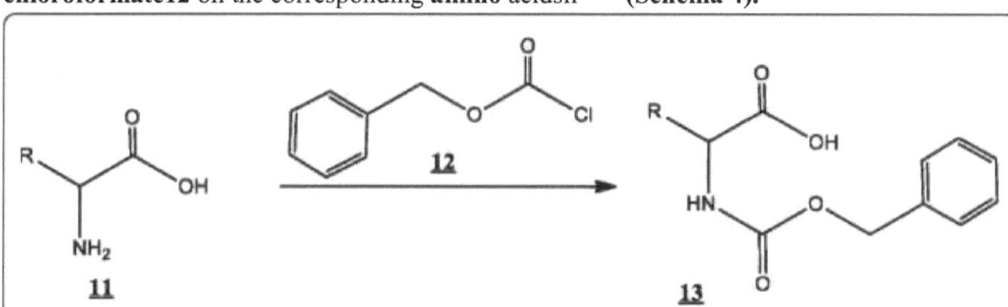

Schéma 4

I.4 Protection by phthalic anhydride (Ft)

Phthalic anhydride is relatively stable under acidic and basic conditions, but can be easily eliminated with nucleophiles. Phthalic acids were first used in peptide synthesis by Kidd and King in 1948[26-27] **(Schema 5).**

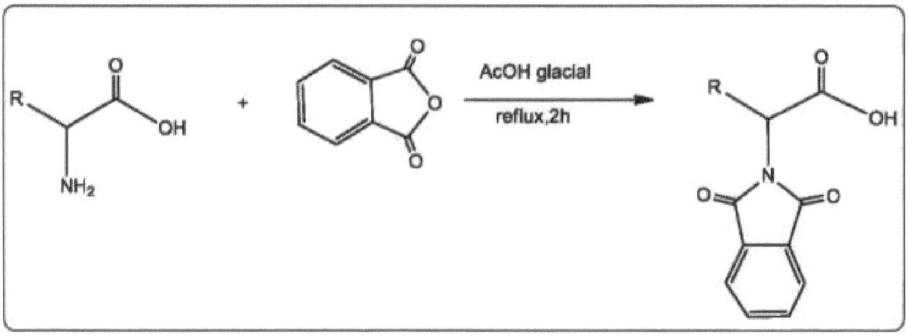

Diagram 5

The mechanism by which phthalic anhydride protects amino acids is as follows **(Figure 6):**

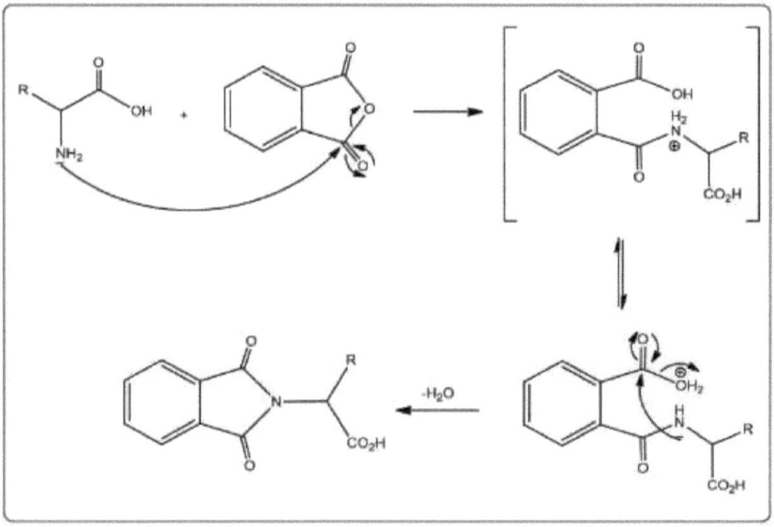

Schema 6

I.5 Protection by tolylsulfonyl or tosyl(Ts)

N-tosyl-a-amino *acids16* can be prepared by the action of tosyl chloride (TsCl)**15** on the corresponding amino-acid **14**. Protection of the amine function is carried out in a sodium hydroxide solution or in a water/acetone mixture in the presence of an equivalent of diisopropylthylamine(iPr)2NEt and an equivalent of sodium hydroxide (NaOH)[28] **(Scheme 7).**

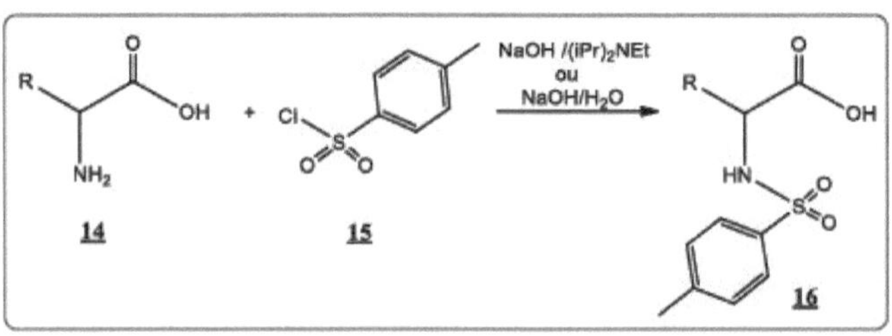

Diagram 7

II. Cleavage of protective groups

The ease with which the amine function can be dëprotëgered determines the choice of protection agent. We give below some mëthodes of deprotection.

II.1 1 Cleavage of t-butyloxycarbonyl (Boc)

To degrade the N-tertiobytoxycarbonyl products, we generally operate in an acid medium[29] **(Schema 8)**:

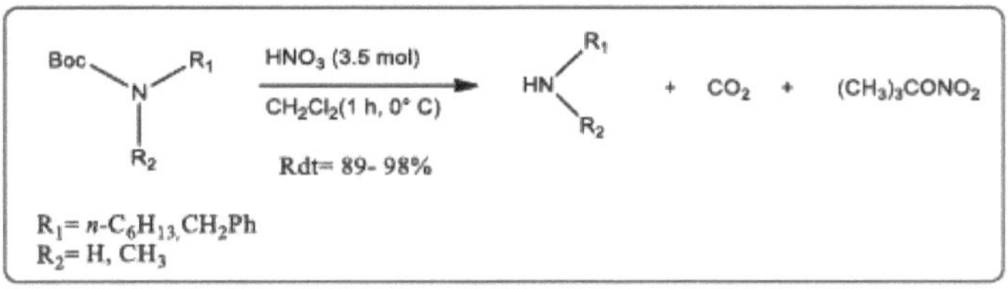

Diagram 8.

In 2004, Norma J. Tom et al[30] developed a process for the deprotection of primary amines protected **17** by the (Boc) group in excellent yields under basic conditions using sodium t-butoxide in tetrahydrofuran or 2- methyltetrahydrofuran (2-MeTHF) **(Figure 9).**

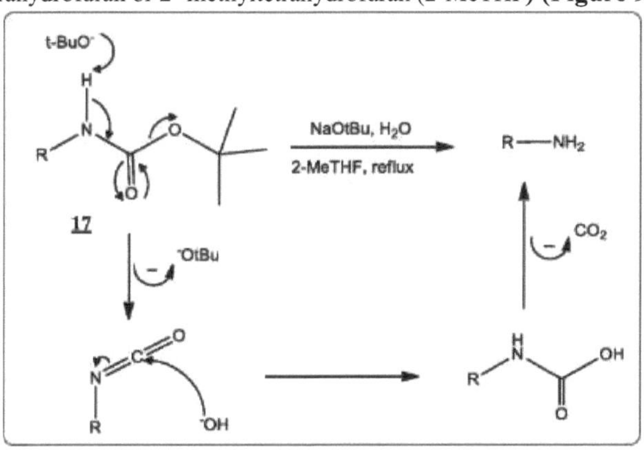

Schéma 9

The dëprotection of t-butyloxycarbonyl in a basic medium can also be carried out by the action of sodium carbonate[31] (Na_2CO_3) in a reflux of dimëthoxyëthane (DME/H2O) **(Schema 10).**

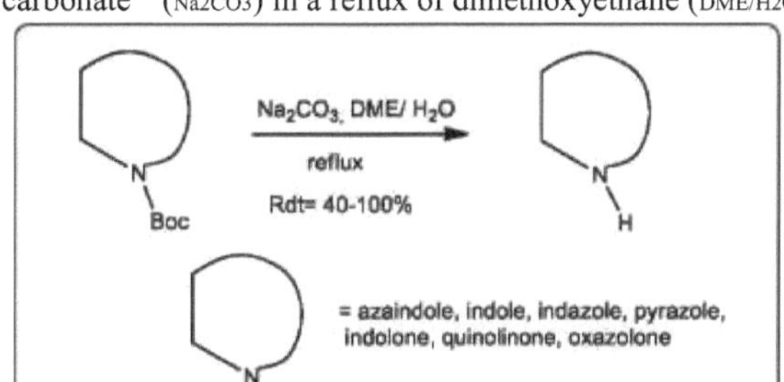

Schéma 10

II.2 2 Cleavage of benzyloxycarbonyl (Cbz)

The dëprotection of the N-benzyloxycarbonyl^s products is rëalisëe gënërally by a catalytic hydrogënation[24-25] **(Schema 11).**

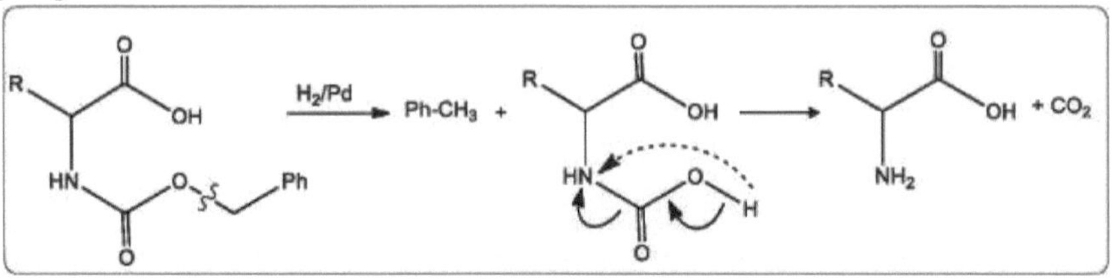

Schema 11

N-benzyloxycarbonate products can also be dëprotëgerated using 10% of the Pd/C and a ëequivalent of sodium tëtrahydшroborate(NaBH4) in шë^ис![32] **(Schema12).**

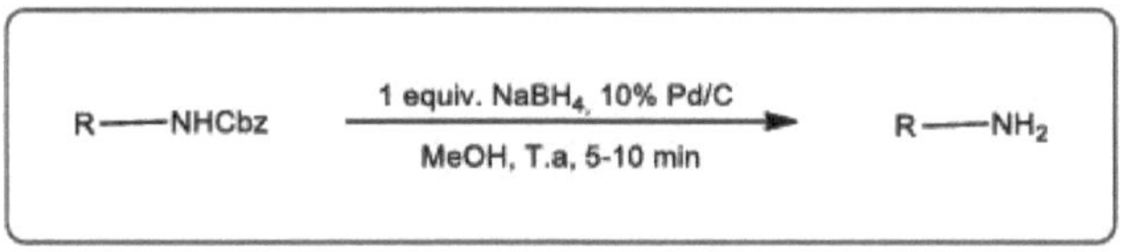

Schema12

II.3 3 Cleavage of phthalic anhydride

The cleavage of phthalic anhydride is often rëalisë by hydrazine in a mëlange of dichloromëthane and mëthanol[26-27] **(Schema 13).**

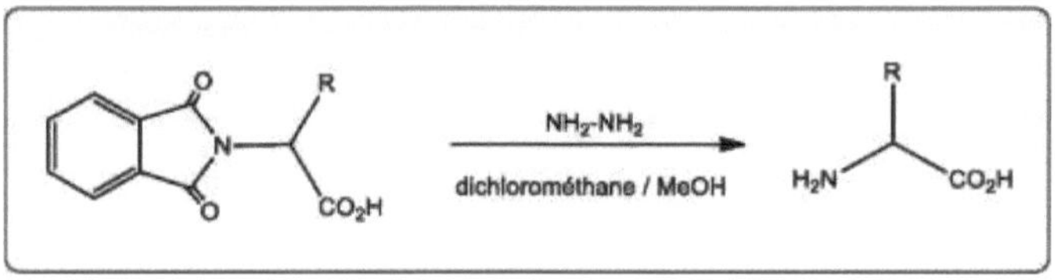

Schema13

II.4 4 Cleavage of tolylsulfonyl or tosyl(Ts)

The protection of N-tosyl products is rëalisëe gënëralement in the presence of a reducing metal such as mag^sium[33] **(Schema 14).**

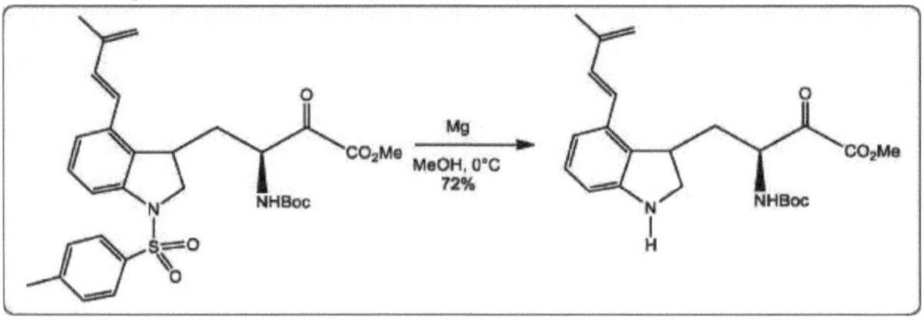

Diagram 14

III. Synthetic interests of protected amino acids

Amino acids monoprotëgës or diprotëgës at the amine function are ШШзёз as a dëpart product for accx'der to a wide range of chiral productssuch asaminocëtones **(18,19and20)**, theN-tosyl-N-mëthylaminoacids21, *theN-* tosylaziridines22andN-tosyl-N-mëthylaminonitriles23 **(Schema15).**

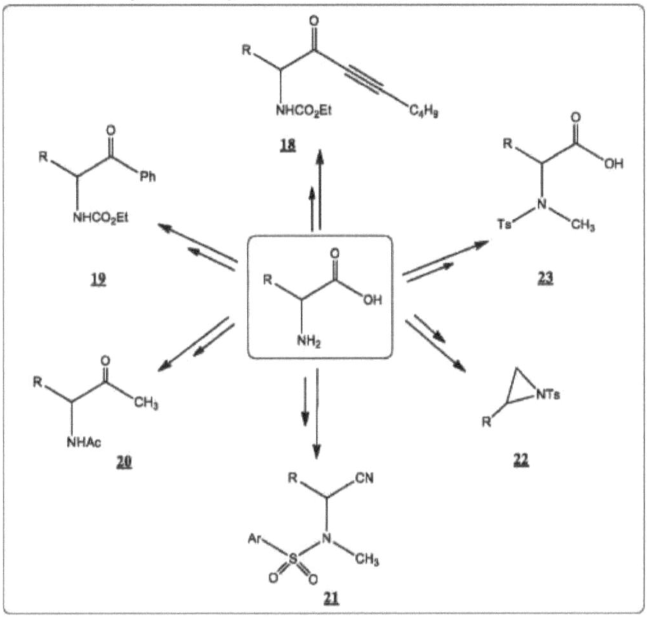

Schémal5

III.1 .1 Synthesis of a-aminocetones

A review of the literature has shown that several methods for synthesizing a-aminocëtones have been developed:

111.1. a-aminoynone synthesis

In 1985, Thomas L. Cupps et al[34] , described a three-step method for the synthesis of **a-aminoynone18** from **N-ëthoxycarbonyl-L-alanine24**. The reaction of product 24with oxalyl chloride in dimëthylformamide (DMF) leads to acid chloride **25.** This intermëdiary is subsequently converted to **pyrrolidide26** by the addition of freshly distilled pyrrolidine, then ^ш^эттë to a-aminoynone18by the action of lithium hexenyltrifluoroborate**(Schema 16).**

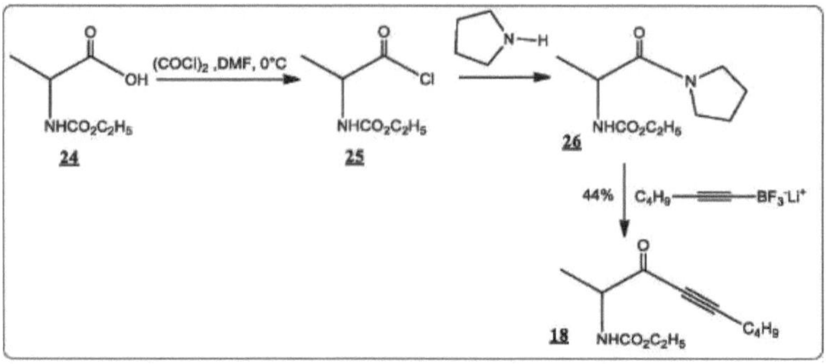

Schéma 16

111.2. b Synthesis of a-aminophenylcetone

This same ëquipe[35] has dëcribed the preparation of a-aminophënylcëtone19by treating **N-ëthoxycarbonyl-L-alanine** *chloride25* prëcëdemment dëcrit with benzëne in the presence of AlChas catalyst[36] **(Schema 17).**

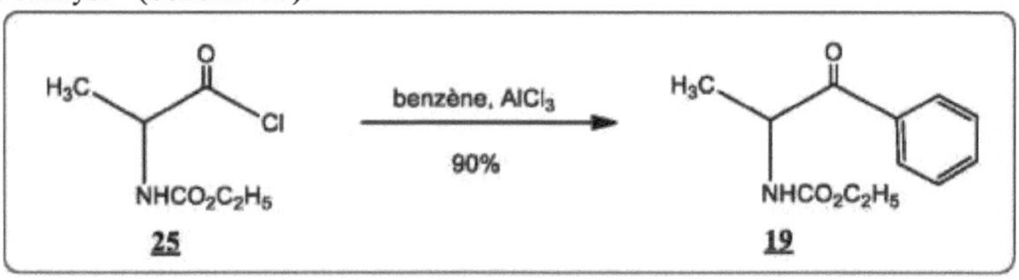

Schéma 17

ПЫ.e Synthesis of a-aminomethylcetone

The direct conversion of an a-amino acid into the corresponding **N-acetyl-a-aminomethylcetone20** by the action of acëtic anhydride (Ac)2o in the presence of pyridine takes place with the release of co2. This reaction is generally known as the "Dakin-West Reaction"[37-39] **(Schema 18).**

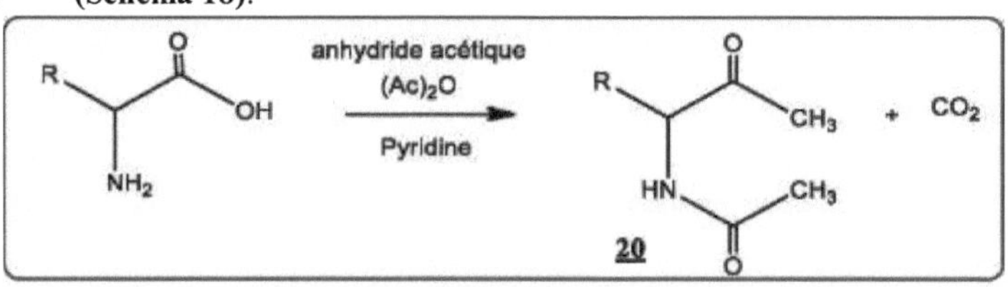

Schéma 18

III.2 .2 Synthesis of optically pure a-aminonitriles

In 2009, our team **at**[40] synthesised new N-methyl-N-arylsulphonyl-a-aminonitriles21 that areomerically pure, starting from the corresponding N-arylsulphonyl-a-amino *acids* 27 via **primary** N-methyl-N-arylsulphonyl-a-aminoamides28.

The use of thionyl chloride as a dehydrating agent gave the following results aminonitriles without any partial racemisation of the asymmetric carbon (**Schema 19).**

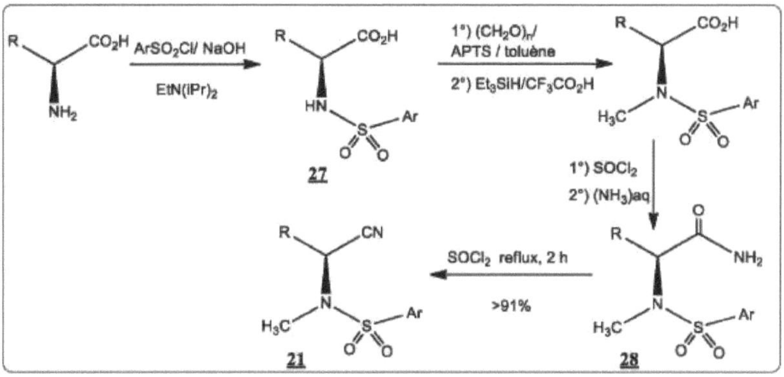

Schéma 19

III.3 Synthesis of aziridines

N-tosylaziridines22were prepared from **N-tosyl29** amino acids, firstly by reduction of the acid function (COOH) forming the corresponding **aminoalcohols30**. The latter is converted to the dësirë 22 product **by** the action of tosyl chloride in the presence of pyridine[41] **(Schema 20).**

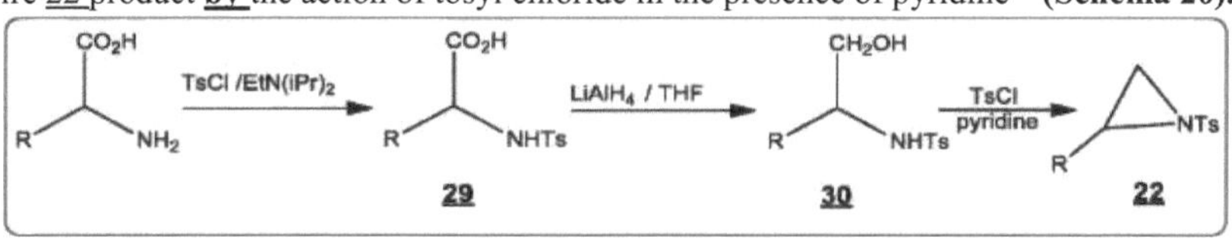

Schema 20

III.4 Synthesis of N-methyl-a-amino *acids*

N-methylamino *acids23* are often used in peptide synthesis.Freidinger et al.[42] , developed a method for synthesising **N-methyl-a-amino** *acids23* which involves condensing **N-protected** amino *acids29* with paraformaldehyde in the presence of paratoluenesulphonic acid (APTS) as a catalyst to form **oxazolidin-5-ones31** which are then reduced by triethylsilane and trifluroacetic acid to give N-tosyl-N-methyl-a-amino *acids23* **(Scheme 21)**. Note that this method is applicable to a variety of amino acids and aldehydes and does not lead to racemisation of the asymmetric carbon.

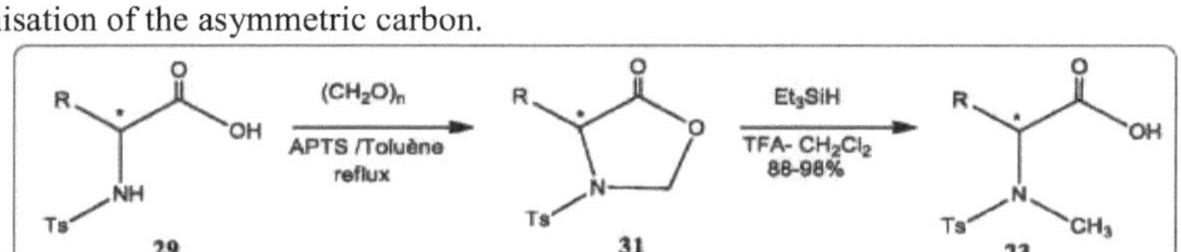

Schéma 21

B. Personal work

I. Synthesis of chiral N-phenylsulfonyl-a-amino *acids*

Our objective in this first part of the work is to protect the amine function of natural amino acids32**(a-f)**. This makes it possible to increase the solubility of our products in ordinary solvents and to work on the acid function without any problems. Protection of the amine

function is carried out with dephenylsulphonyl chloride in the presence of an equivalent of sodium hydroxide and an equivalent of diisopropylethylamine in a water/acetone mixture (1/1) at room temperature.
The synthesis of N-phdnylsulfonyl-a-amino *acids33*<u>**(a-f)**</u>is illustrated in the following reaction schemes **(Schemes 22 and 23).**

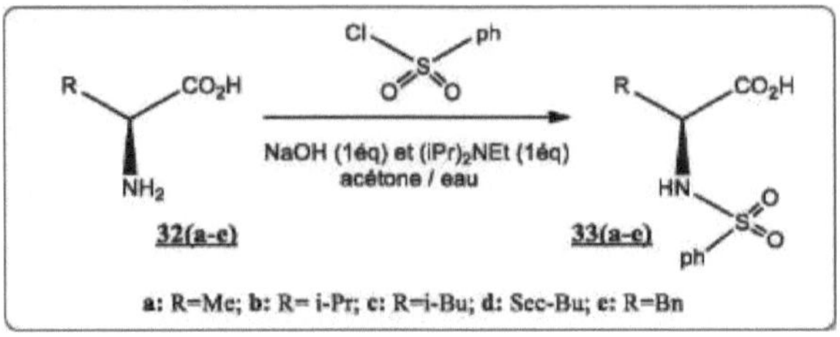

Diagram 22

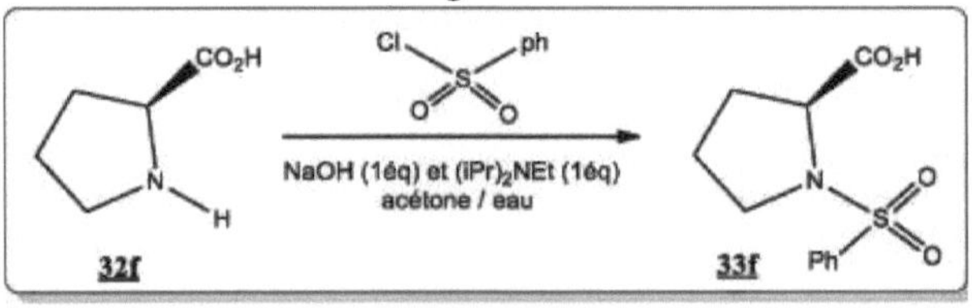

<u>**Schéma 23**</u>

Mécanisme réactionnel

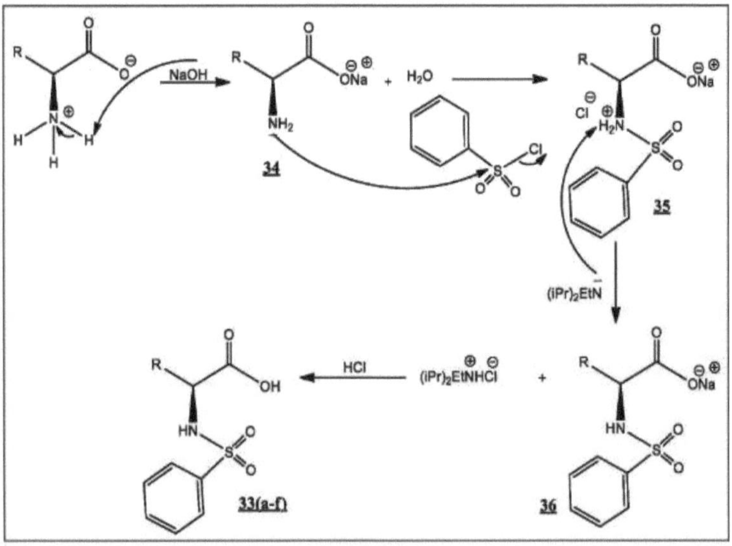

<u>**Schéma 24**</u>

The mechanism starts with a dëprotonation of the amine function by sodium hydroxide to form the **salt34**. The intermëdiary 35 **<u>is</u>** obtained **<u>by</u>** nucleophilic attack of the amine on the sulphur of phënylsulphonyl chloride. The addition of a base such as diisopropylethylamine leads to the N-phenylsulfonyl acid *salt36 which* is converted to the dësirë product by the

addition of an excës of hydrochloric acidconcentrate **(Schema 24).**
The chemical yields and physical aspects of the N-phenylsulfonyl-a-amino *acids* prepared are shown in the table below:

Table I

Chemical yields and physical aspects of prepared N-phenylsulfonyl-a-amino *acids*

Amino acids	R	Yield % (%)	Aspect	Melting point (°C) ± 2°
Alanine	Me	93	Solid white	132
Valine	i-Pr	80	Solid white	138
Leucine	i-Bu	85	Solid white	122
Isoleucine	Sec-Bu	83	Solid white	142
Phenylalanine	Bn	91	Solid white	160
Proline	-$(CH_2)_3$-	80	Solid white	58

II. Spectroscopic studies of N-phenylsulfonyl-a-amino *acids*

The N-phenylsulphonyl-a-amino *acids* prepared in this work were identified by H and C NMR at[1] and[13] .

II.1 1 1H NMR spectroscopic properties

The 1H NMR spectroscopic study confirmed the structures of the *N*- phenylsulphonyl amino acids prepared (see experimental section). As an example, let's look at the case of *N*-phenylsulphonyl-(L)-**valine33b**:

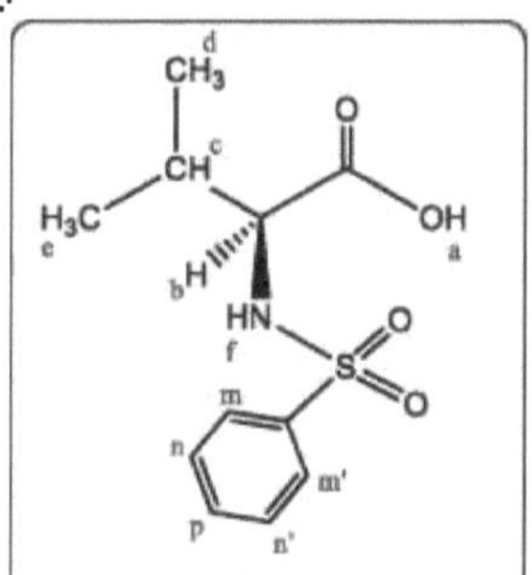

33b

***N*- phenylsulfonyl-(L)-valine**

The 1H NMR spectrum of **composë33b(Figure 11)**recorded at 300 MHz in deutërë chloroform shows the following signals:

- A 3H integration doublet has 0.88 ppm corresponding to the mëthylic protons $(CH_3)_e$or$(CH_3)_d$ with a coupling constant (J= 6.9 Hz).
- A 3H inegration doublet at 0.96 ppm corresponding to the mëthylic protons $(CH_3)_d$ou $(CH_3)_e$ with a coupling constant (J= 6.6 Hz).
- A proton inegration multiplet between 2.04 and 2.16 ppm corresponding to the H_c proton.
- A doublet dëdoublë inegration a proton at 3.81 ppm corresponding to the Hb proton (J=9.9 Hz; J=4.8 Hz).
- An intëgration doublet a proton at 5.15 ppm corresponding to the Hf proton with a coupling constant (J=9.9 Hz).

- A multiplet of 3H intëgration between 7.45-7.58 ppm corresponding to the Hn,Hn' and Hp protons.
- A 2H intëgration doublet has 7.85 ppm corresponding to protons Hm and Hm' with a coupling constant (J=7.8 Hz).

II.2 2 NMR spectroscopic properties C[13]

The full proton decoupling NMR spectrum[13] C of the same **compse33b (figure 12)** recorded in deuterated chloroform at 75 MHz confirms the structure based on the following signals:

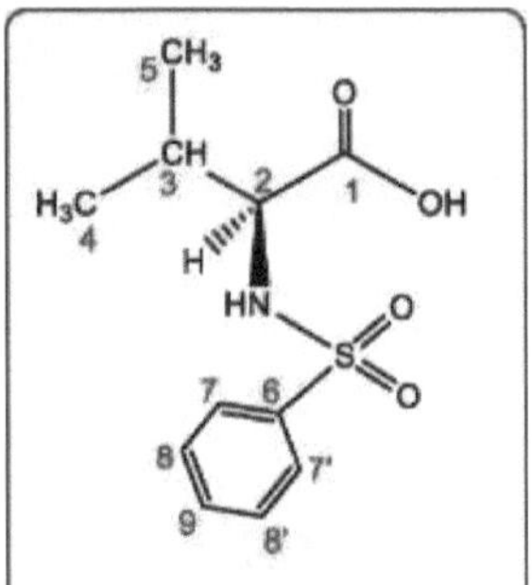

33b

- A signal at 16.69 ppm corresponding to carbon C4OU C5.
- A signal at 18.36 ppm corresponds to carbon C5 or C4.
- A signal at 30.94ppm corresponds to carbon C3.
- A signal at 60.18ppm corresponds to carbon C2.
- Signals at 126.73; 128.49; 132.38; 139.32 ppm corresponding respectively to the aromatic carbons C8.8'; C7.7'; C9; C6.
- A signal at 174.70 ppm corresponds to the C1 carbonyl.

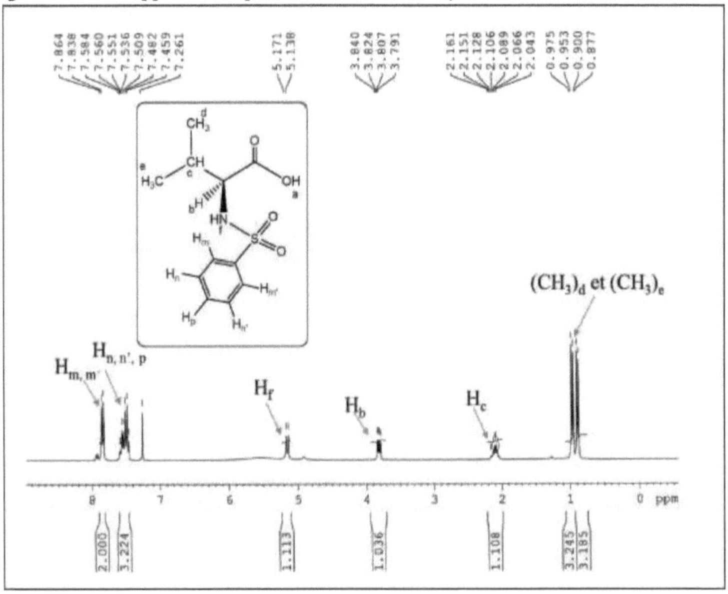

Figure 11. H NMR spectrum of compound 33 recorded at 300 MHz (CDCl)

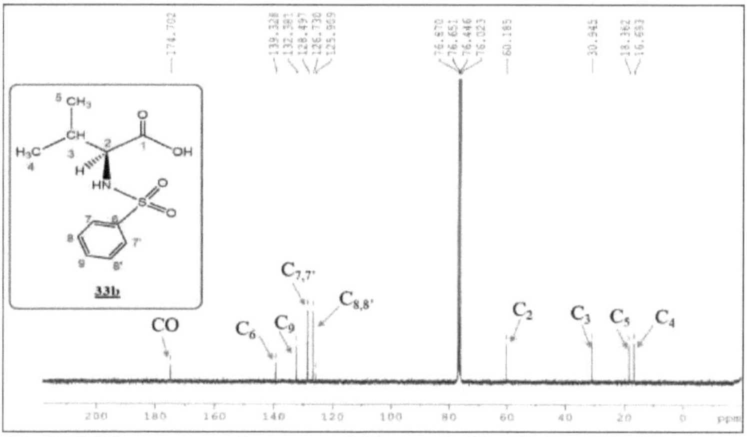

Figure 12. C NMR spectrum of compound 33 recorded in CDCl at 75 MHz.

C. Conclusion

In this first chapter, we have rëalisëed the synthesis of a sëerie of **33(a-f)**chiral *N*-phenylsulfonyl-a-amino acids in yields ranging from 80% to 93%. This protection is rëalisëe by treating the natural a-amino acids with phënylsulfonyl chloride. The monoprotëgëchiral a-amino acids obtained will be used to prepare corresponding a-amino esters, which will be the subject of the next chapter.

Chapter II

Synthesis *of* N-phënylsulfonyl-a-ammoesters chiral

A. General information and bibliographical references

The literature review mo nt that a-aminoesters are important intermediates that have ë1.ë и^^ë in peptide synthesis[43] and medicinal chemistry .[44-45]

In this chapter, we will describe some significant examples of the use of a-aminoesters in organic synthesis, as well as their mëthods of preparation, and we will then present our personal work.

I. Synthetic interests

A wide range of products of biological interest have ëlë been prepared using a-aminoesters. Some examples are given below:

I.1 Synthesis of 1,2,4-oxadiazin-6-ones by condensation of aromatic nitrile oxides with a-amino esters

The first synthesis of 1,2,4-oxadiazin-6-one was carried out by Hussein et al.[46] , in 1984. This method consists of condensing **aromatic** nitrile oxides37 with a-amino ester **types38** to give **intermediates39** which cyclize to give 1,2,**4-oxadiazin-6-ones40** in chemical yields of between 12% and 73% **(Schema 25)**.

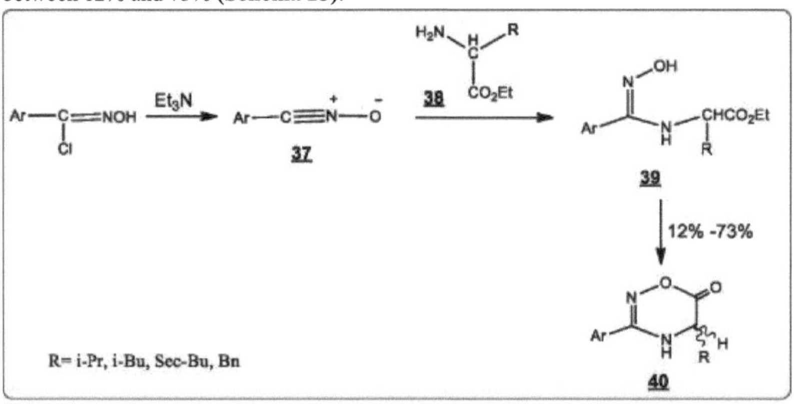

Schéma 25

I.2 Synthesis of an azabenazoline derivative from glycine methyl ester

In 2010, C. Lamberth[47] synthëtisë an auxin herbicide in three ëtages from glycine mëthyl **ester41**. This is treated with carbon disulphide (CS2) and sodium hydroxide to give the **unstable** dithiocarbamate42. The addition of chloroacëtonitrile to the reaction mixture leads to the formation of 4-imino-2-thioxo-3-thiazolidine acetate43. This intermediate is condensed with в-ethoxyenone in the presence of a catalytic amount of piperidine to give a thiazolo [4,5-b] pyridine **derivative44**, which is converted by desulphurisation with the enderived mercuric trifluoroacetate of azabenazoline45**(Scheme 26)**.

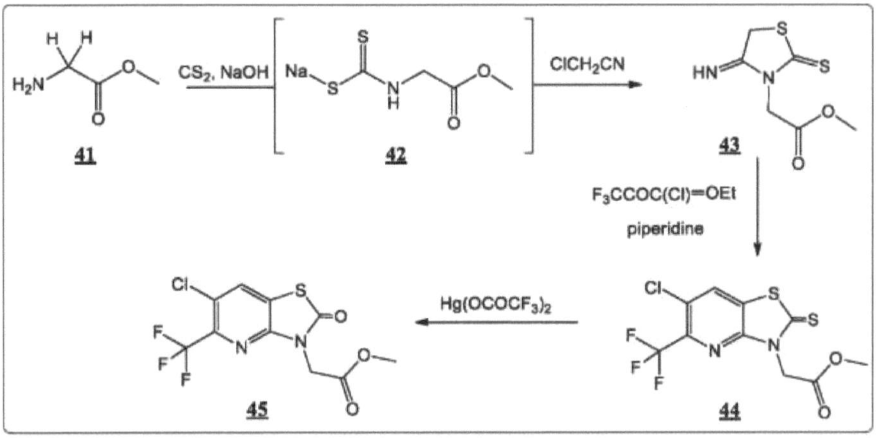

Schéma 26

I.3 Synthesis of fenamidone from (S)-methylphenylglycine methyl ester
In 2010, C.Lamberth also synthesised fenamidone[47] in three steps from (S)-methylphenylglycine methyl **ester46**. The first step consists of adding lathiophosgene to the methyl ester to give the **isothiocyanate47** which is then converted to thiohydantoin48 by the addition of phenylhydrazine. Finally, the desired product [(5S)- 5-methyl-2-(methylsulfonyl)-5-phenyl-3-(phenylamino)-3,5-dihydro-4H-imidazole-4-one] 49is obtained by sulfur methylation **(Scheme 27).**

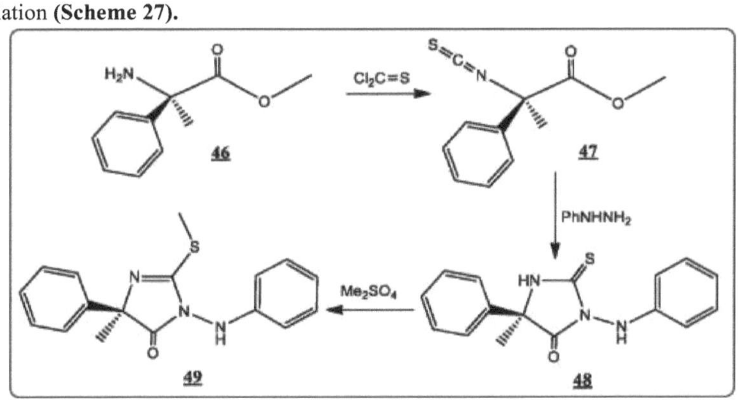

Schema 27

I.4 Synthesis of new chiral ligands from methyl 2-amino-3-methylbutanoate
U.balakrishnan et al,[48] have rëalisës synthesized new chiral ligand from тёЛук **50** 2-amino-3-mëthylbutanoate. Condensation of the latter with 5-chloro-2- **hydroxybenzaldehyde51** under reflux of toluene leads to Гmtermëdiary **52** which subsequently gives the dësirë product **53** by reduction in the presence of mëthanol **(Scheme 28)**.The product 53aëtë ШШзё as a chiral ligand to catalyse ënantiosëlective reduction reactions of cëtones **(Scheme 29).**

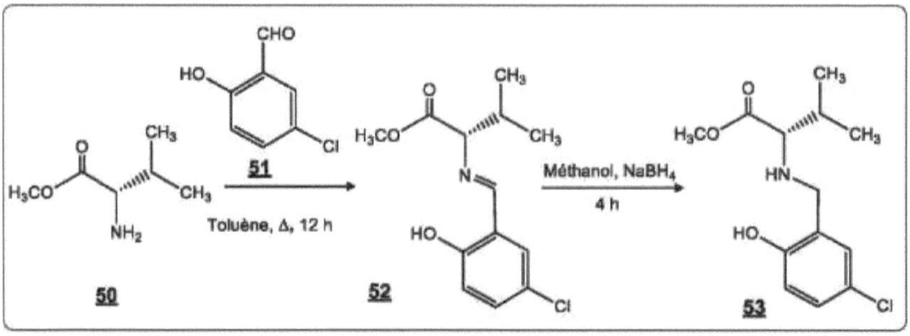

Schéma28

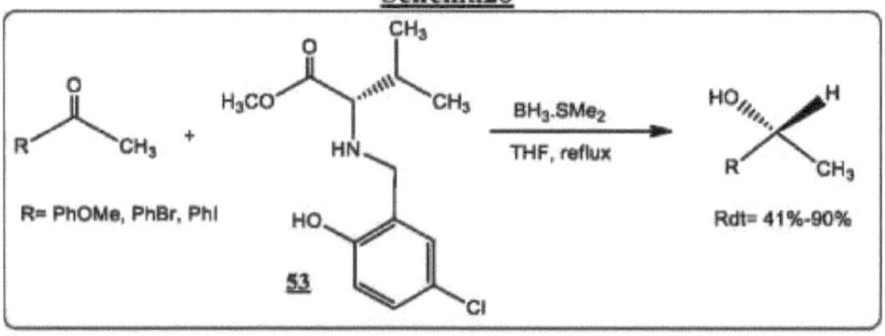

Schéma 29

II. Different methods for preparing a-amino esters

The literature review showed that acid esterification can be achieved by several methods.

II.1 1 Esterification in an acid medium

C.Lamberth[47] synthesised (S)-methylphenylglycine methyl **ester46** by adding methanol to (S)-**methylphenylglycine54** in an acid medium **(Schema 30)**.

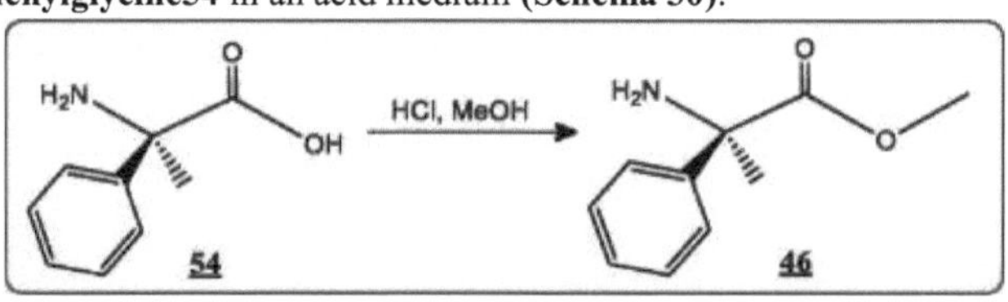

Schema 30

II.2 2 Esterification in a basic medium

P. Strazzolini et al 29 synthesised benzyl **esters56** from *N-Boc*(L)-amino **acids55**. Treatment of the corresponding protected a-amino acids with 1'a-bromo toluene (PhCH2Br) in the presence of 1'sodium hydride in 1,1'-oxybis[2-methoxyethane] (diglyme) afforded **benzylesters56** in chemical yields of between 47% and 72% **(Scheme 31).**

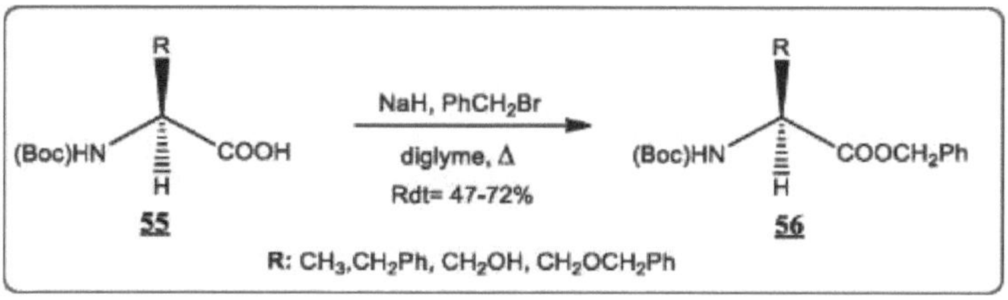

Schema 31

II.3 3 Esterification with chlorotrimethylsilane (TMSCl) and methanol

Jiabo Li and yaowu Sha[49] have prëparë a sërie of aminoesters тё^уНдие5^о^ form of a hydrochloride salt by the reaction of amino acids57non protëgës with anhydrous nrethanoland chlorotrimëthylsilane (TMSCl) (**Schema 32).**

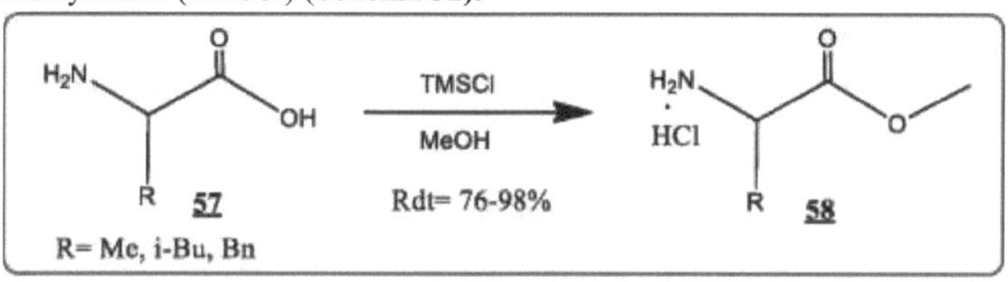

Schéma 32

The use of TMSCl / MeOH has the following advantages: operation is easy, conditions are mild, the reaction is simple and yields are good to excellent.

II.4 4 Esterification with methanol in the presence of thionyl chloride (SOCh)

U.Balakrishnan et al,[48] synthëtisësle (L)-valine mëthyl ester60 in the form of a hydrochloride se l by the reaction of (L)-valine 59 with methanol and thionyl chloride (SOCh) at room tempërature (**Scheme 33).**

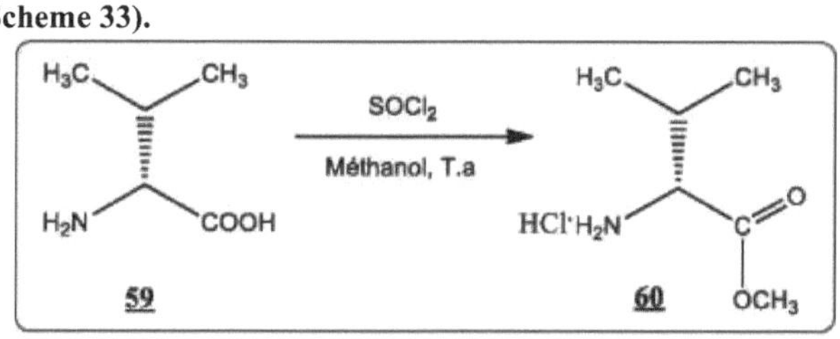

Schéma 33

B. Personal work

1. Synthesis of chiral N-phenylsulfonyl-a-amino *esters* from N-phenylsulfonyl amino acids

In this work, we adoptedë the mëthod of Jiabo Li and yaowu Sha,[49] rëalisëe on unproëgësed amino acids, to transformë the N-phenylsulfonyl amino acids prepared in the first chapter into the corresponding a-aminoesters.

Note that in the case of free amino acids, the esters are obtained in the form of hydrochloride salts.

Treatment of N-protected a-amino acids **33(a-f)**with chlorotrimethylsilane and methanol at room temperature afforded the a-amino *esters* N-phenylsulphonyls61**(a-f)**in good chemical yields **(Schemes 34 and 35).**

The reaction sequence is summarised in the following diagram:

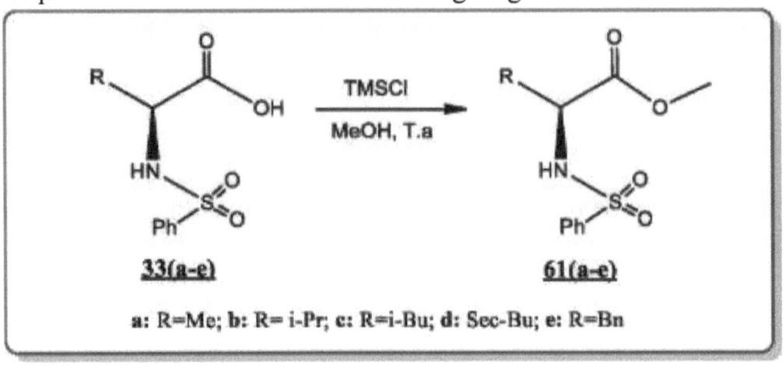

Schema 34

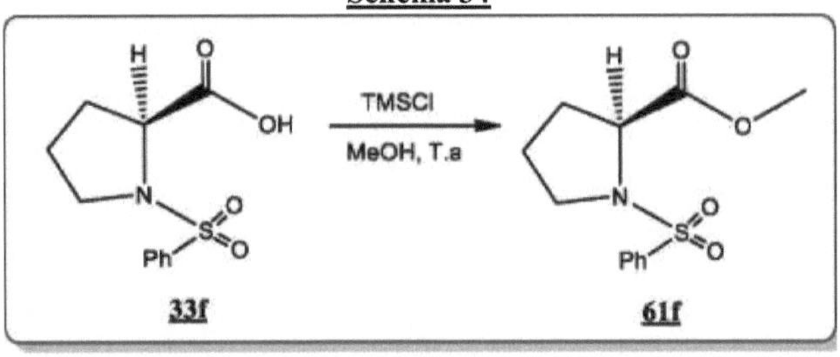

Schéma 35

The mechanism of this reaction is as follows **(Schema 36)**:

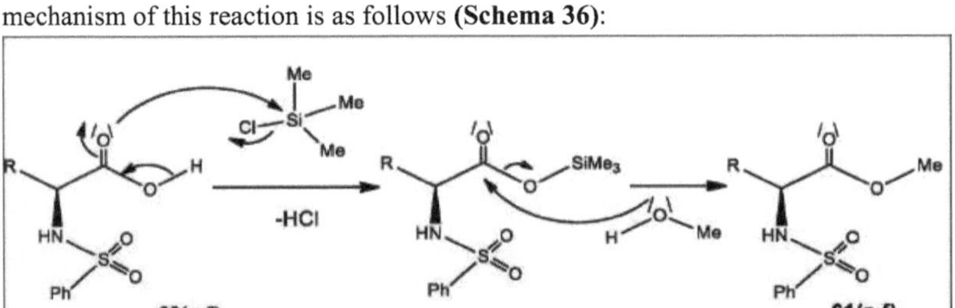

Schéma 36

The chemical yields and physical aspects of the products prepared are given in

the following table:

Table II

Chemical yields and physical aspects of prepared a-amino esters61 **(a-f)**

Compose	R	Yield	Aspect	Melting point (°C) ± 2°
61a	Me	94	Solid white	66

61b	i-Pr	74	Solid white	92
61c	i-Bu	95	Solid white	70
61d	Sec-Bu	88	Solid white	68
61e	Bn	95	Solid white	104
61f	-(CH2)3-	95	Solid white	86

11. Spectroscopic studies of prepared N-phenylsulfonyl-a-amino *esters*

The a-aminoesters were identified by NMR spectroscopy at[1] H and C.[13]

11.1 1H NMR spectroscopic properties

NMR spectroscopy[1] H confirmed the structures of the N-phenylsulfonyl-a- aminoesters prepared (see expërimental section). As an example, let us study the case of **methyl61d** N-phenylsulfonyl-2-amino-3-methylpentanoate **(Figure 13).**

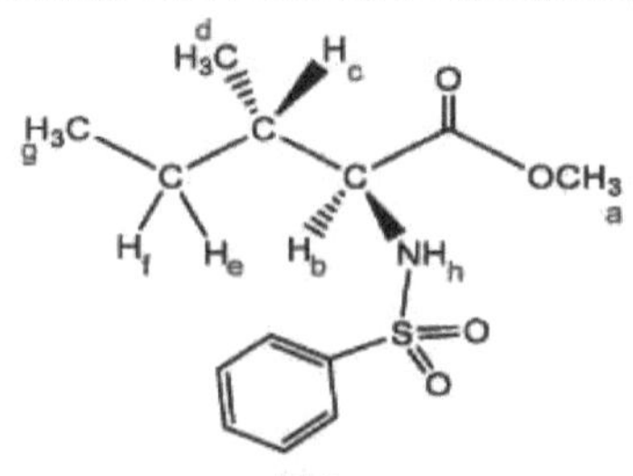

61d

N-phenylsulfonyl-Z-amino-S-methylpentanoatedemethyl

The 1H NMR spectrum of **compose61d (Figure 13)** recorded at 300 MHz in deuterated chloroform shows the following signals:

- A multiplet between 0.84-0.90 ppm 6H integration related to the methyl protons (CH3)det(CH3)g.
- A multipletre1 .10-1.20 ppm of proton integration relative to the Hfou He proton.
- A multipletre 1 .38-1.46 ppm of integration of a proton relative to the He or Hf proton.
- A multipletre 1 .70-1.80 ppm of integration of a proton relative to the Hc proton.
- A singlet has 2. 16ppm of 3H integration relative to the protons of the methoxy group. (OCH3).
- A doublet doubles at 3.79 ppm integration of a proton relative to the H proton with a coupling constant (J=5.7 Hz; J=9.9 Hz).
- A doublet has an integration of 5.14 ppm, a proton relative to the Hh proton with a coupling constant (J=9.9 Hz).
- A multiplet between 7.46-7.56 ppm 3H integration relating to protons Hn,Hn' and Hp.
- A doublet has 7.83ppm of 2H integration relative to protons Hm and Hm' with a coupling constant (J=7.5 Hz).

11.2 NMR spectroscopic properties C[13]

The full proton decoupling NMR spectrum[13] C of the same compound **61d (Figure 14)** recorded in deuterated chloroform at 75 MHz confirms the structure based on the following signals:

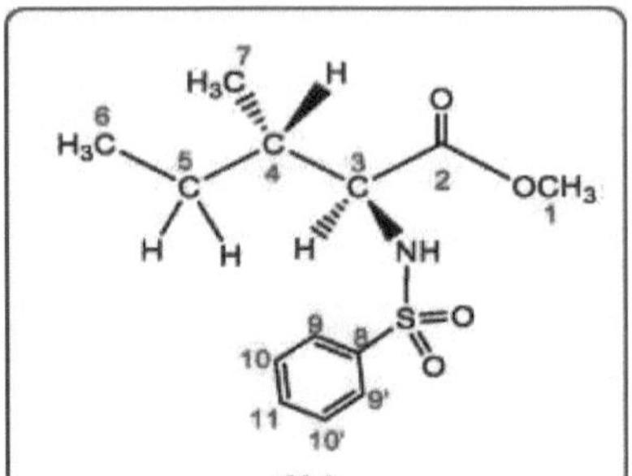

61d

- A10 . 69ppm signal for carbon C6.
- A 14 .82ppm signal for carbon C7.
- A signala24 .15ppm relating to carbon C5.
- A 37 .90ppm signal for carbonC4.
- A51 .60ppm signal for carbonCi.
- A 59. 77ppm signal relating to carbonC3.
- Signals at 126.79; 128.47; 132.27; 139.11 ppm corresponding respectively to the aromatic carbons C10,10'; C9,9'; C11; C8.
- A signal at 171.17 ppm related to carbonyl C2.

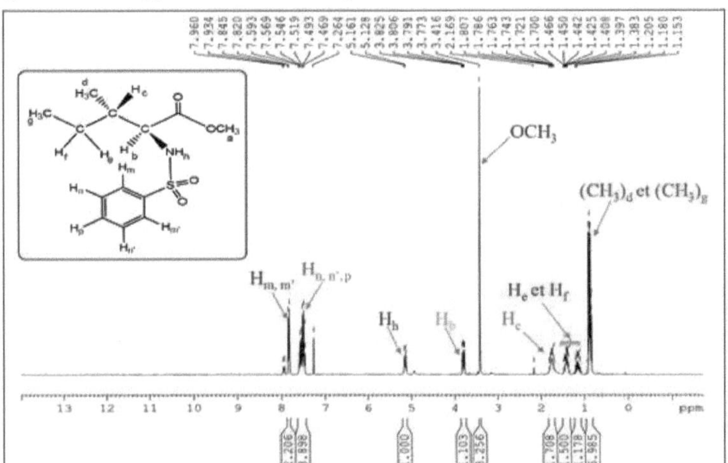

Figure 13. H NMR spectrum of compound 61d recorded at 300 MHz (CDCl)

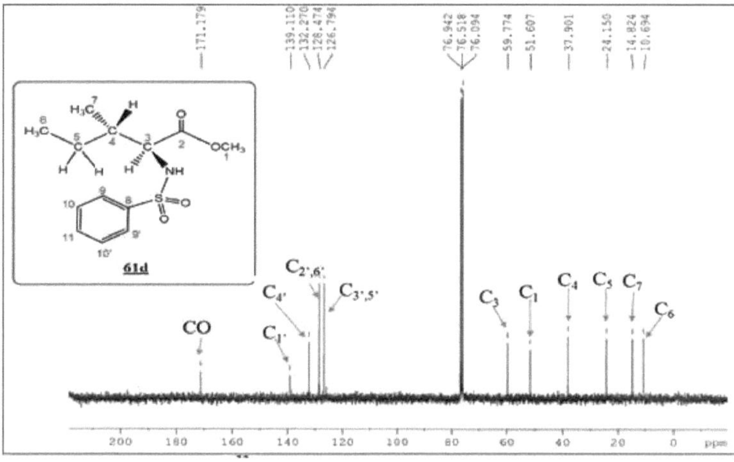

Figure 14. C NMR spectrum of compound 61d recorded in CDCl at 75 MHz
3

C. Conclusion

During this second chapter,we rëalisë the synthesis of a sërie of six *a*- aminoestersN-pЬёпуки1£опу^61**(a-0** chiral with chemical yields between 74% and 95%. The conversion of these esters to the corresponding hydrazides will be the subject of the next chapter.

Chapter III

Synthesis of *chiral* N-phënylsulfonyl-a-ammohydrazides chiral

A. General information and bibliographical references

A review of the literature shows that hydrazides are important precursors for the synthesis of hëtërocycles[50-54] , pharmaceuticals, polymëers and dyes[55] . In particular, chiral hydrazides are soughtës in the field of asymmetric synthesis and several teams have attempted to prepare them in enantiomerically pure form.

In this chapter we will first give an overview of the different methods of synthesising hydrazides and their applications. We will then present our personal work.

I. Synthetic interests of hydrazides

Hydrazides are highly reactive compounds that have been used as starting materials for new heterocyclic structures. As an example, type 62 hydrazide was reacted with pentane-2,4-dione, ethoxyalkylidene and phenylthioisocyanate respectively to give the products[56] **(63,64,65)(Schema 37).**

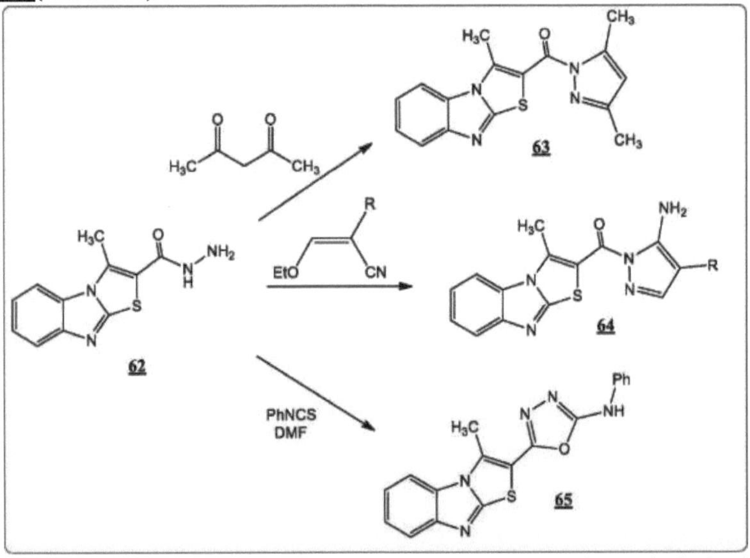

In what follows, we give more examples of the application of a-aminohydrazides in hëtërocyclic synthesis.

I.1 Synthesis of 1,3,4-oxadiazoles from a-aminohydrazides

1,3,4-oxadiazoles play a particularly important role in medicinal chemistry due to their various biological activities: anti-tumour[57-59] , antibacterial[60] and antimicrobial[61-64] . In this context, kudelko et al,[50] have developed a new method for the synthesis of 2-aminomethyl-1,3,4-oxadiazoles. Treatment of **a-aminohydrazides66** with triethylorthoesters in benzene reflux and in the presence of glacial acetic acid led to the formation of 2,5-disubstituted-1,3,4-**oxadiazoles67** in good chemical yields **(Schema 38).**

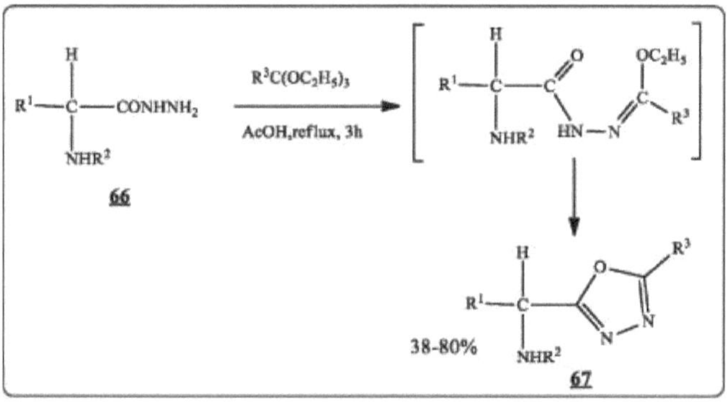

Schéma 38

I.2 Synthesis of salicylidene hydrazide

In 2006, J. Becher et al.[65] developed a synthesis route for the first carbohydrate derivative salicylidene **hydrazide70** which represents an important class of chiral ligand in oxo vanadium complexes. This synthesis was carried out by condensation of hydrazide68withhydroxybenzaldehyde (salicylaldehyde)**69in** refluxing ethanol for six hours **(Scheme 39).**

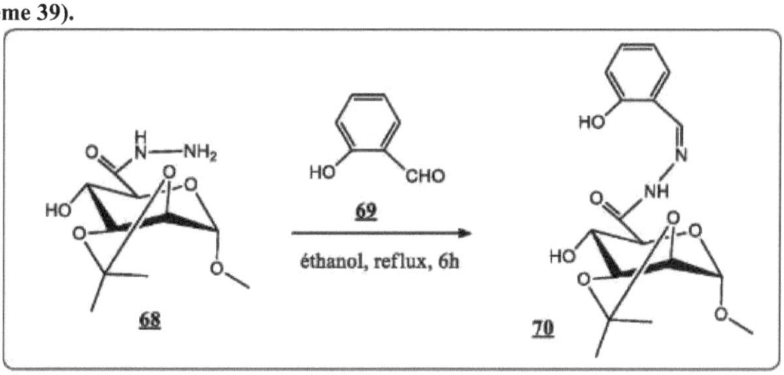

Schéma 39

It should be noted that hydrazide **68** was prepared from mCthyl-a-(D)**-mannopyranoside71** via three steps as follows:

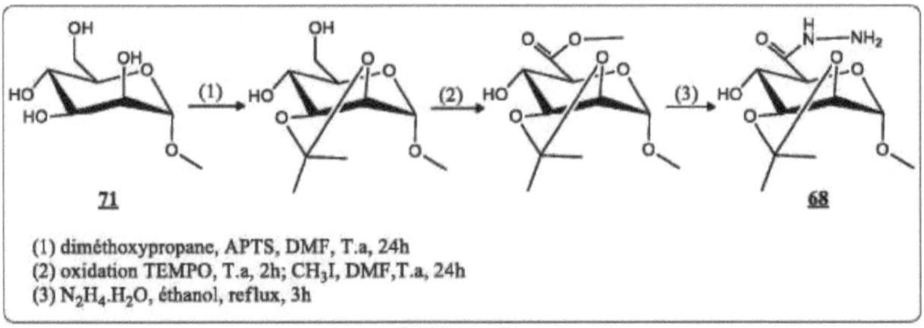

Schéma 40

I.3 Synthesis of *4-acetyl-N-benzylidene-3*,5-dimethyl-1H-pyrrole-2-carbohydrazides

In the search for new pulmonary anti-tuberculosis agents, AtulManvar et al.[66] synthesised *4-acCtyl-N-benzylidene-3*,**5-dimCthyl-1H-pyrrole-2-carbohydrazides73** from pyrrole-hydrazide72 using glacial acdtic acid as a catalyst **(Schema 41).**

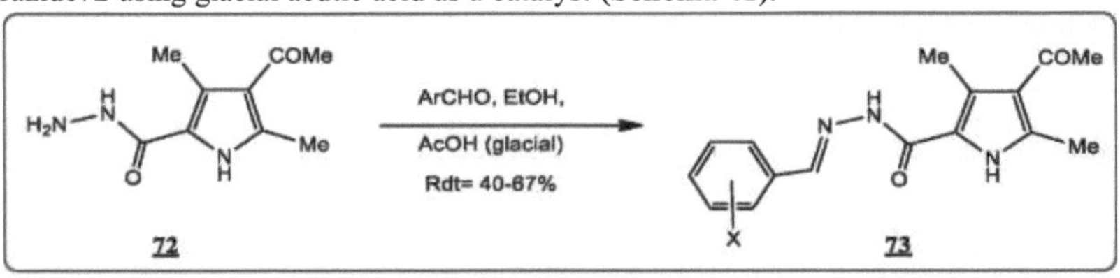

Schéma 41

I.4 sandwich structure of commercial ruthenium porphyrin and valine hydrazide

In 2011, Q.-F. Liang et al[67] rëalisë the sandwich structure of commercial ruthenium porphyrin [RuII(TDCPP)CO] and lavaline **hydrazide74**. These complexes are usedës to derivatise amino acids and dëterminate the absolute configuration based on NMR and circular dichroism **(Schema 42).**

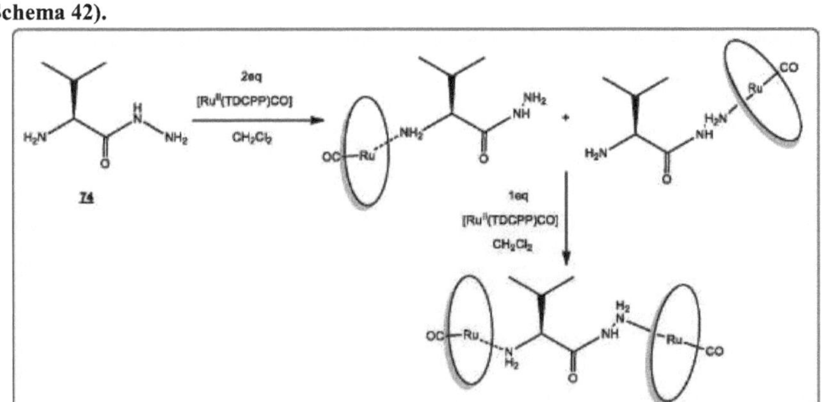

Schema 42

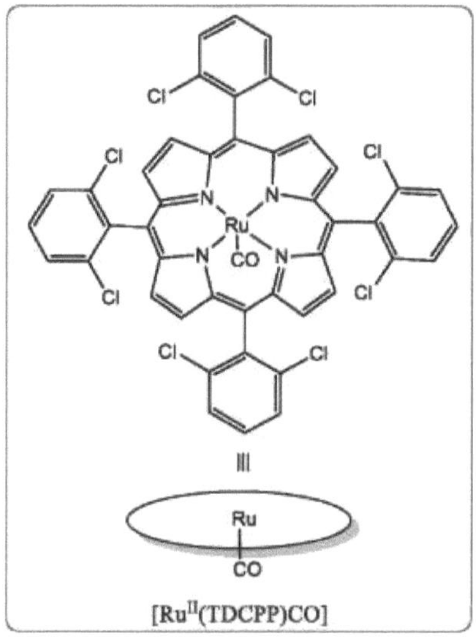

[Ru^{II}(TDCPP)CO]

Figure 15: sandwich structure of commercial Ruthenium porphyrin

II. Main routes for preparing hydrazides

The methods for preparing hydrazides are very varied. In the following we describe the main methods cited in the literature.

II.1 1 Synthesis of a-aminohydrazides from a-amino acids

Q-F. Liang et al.[67] , have described the two-step synthesis of a-amino hydrazide from (L)-valine. The first step consists of condensing (L)-**valine75** with thionyl chloride in refluxing methanol to give the **type76** amino ester. This intermediary is transformed directly into **aminohydrazide77** by the action of hydrazine hydrate in methanol at room temperature **(Schema 43).**

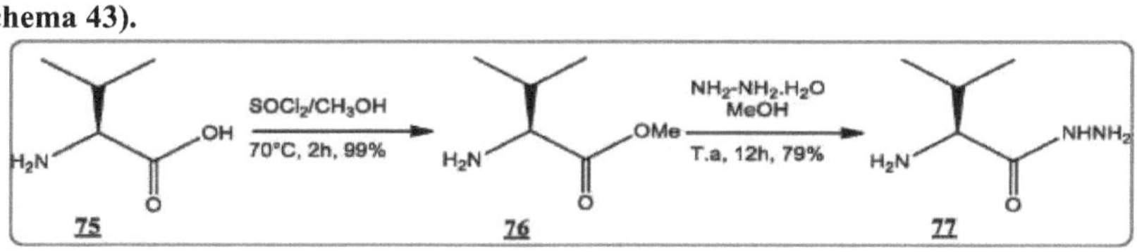

Schema 43

II.2 2 Synthesis of hydrazides by reduction of hydrazones

D. Perdicchia et al.[68] , (E)-2-benzilidene-1,1-dimethylhydrazine 79 was synthesised by condensation of **dimCthylhydrazine78** with benzaldehyde (PhCHO) in xylenea Compound **79** yields intermediate **80** by addition of borane and trimCthylamine in an acid medium. It is then transformed into the product dCsirC by acylation**(Schema 44).**

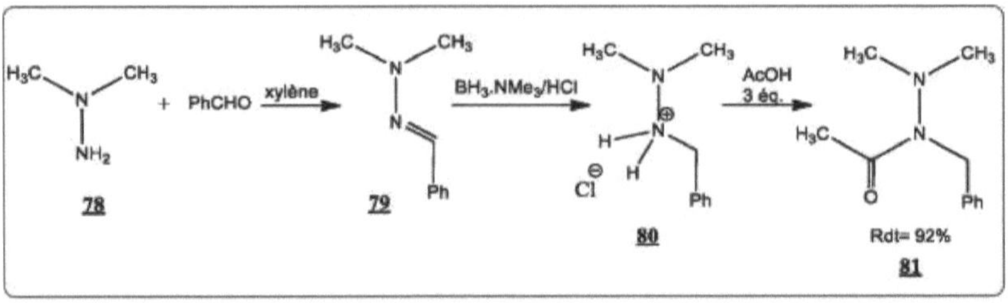

Schéma 44

II.3 3 Synthesis of a-mercaptohydrazides from a-bromoacids

Lesa-mercaptohydrazides ont ël.ë obtenus *a* partir des a-bromoacidess selon une mëthode dëcrite par Kudelko[69] Les **a-bromoacides82** sont traités avec une solution aqueuse deNaSH puis esterifie avec du methanol pour donner les **thioesters84**. The latter is condensed with hydrazine hydrate to form the corresponding hydrazides **85** in good chemical yields **(Schema 45).**

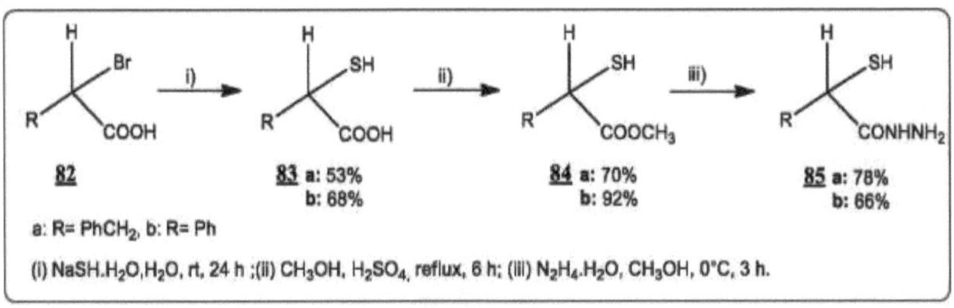

Schéma 45

II.4 4 Preparation of a-aminohydrazide from (s)-leucine

Isaac N. Arthur et al[70] achieved the synthesis of a-aminohydrazide from (s)-**leucine86** in four steps. The first step consists of protecting the amine functions followed by estërification of the acid. The addition of N-bromosuccinimide(NBS) results in the **intermëdiary87** which reacts with silver fluoride (AgF) to give the **intermëdiary88**. Finally, the latter is converted to the corresponding aminohydrazide 89by the addition of hydrazine **(Schema 46).**

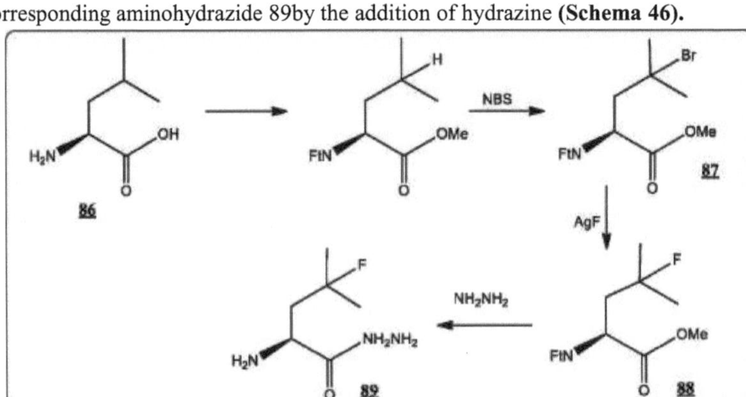

B. Personal work

1. Synthesis of chiral N-phenylsulfonyl-a-aminohydrazides from N-phenylsulfonyl-a-aminoesters

The synthesis of a-aminohydrazides90**(a-f)** is illustrated in **(Schemes 47 and 48)**. N-phenylsulphonyl-a-aminoestes61**(a-f)**react with 1'hydrazine in refluxing methanol to give theiratical a-aminohydrazides.

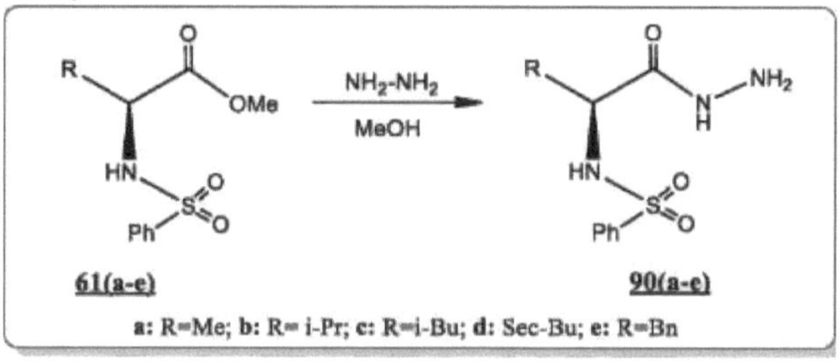

Schéma 47

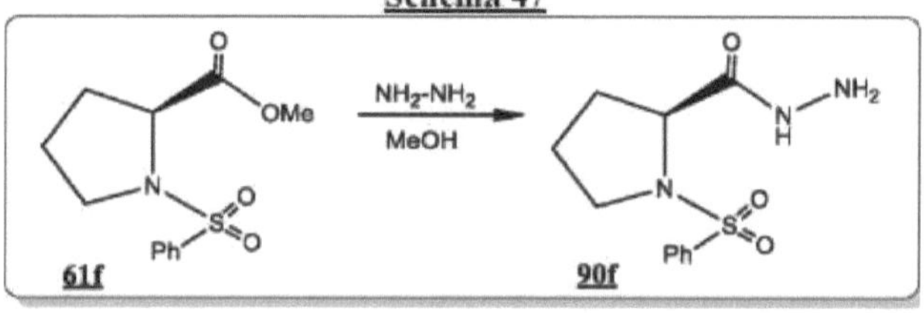

Schéma 48

Reaction mechanism

The mechanism consists of a nucleophilic attack of the free doublet of the hydrazine nitrogen on the carbonyl of the ester function and the departure of the methoxy group **(Schema 49).**

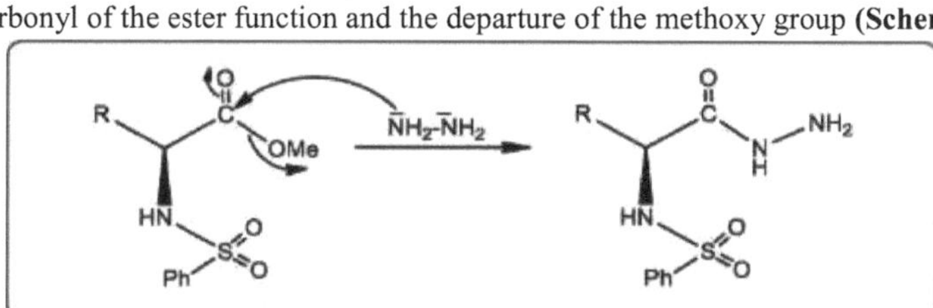

The chemical yields and physical aspects of the products prepared are given in the table below:

Table III

Chemical yields and physical aspects of prepared a-aminohydrazides

composes	R	Yield	Aspect	Melting point (°C)
90a	Me	76	Solid white	84
90b	i-Pr	87	Solid white	150
90c	i-Bu	92	Solid white	92
90d	Sec-Bu	81	Solid white	136
90e	Bn	78	Solid white	159

90f	-(CH2)3-	85	Yellowish oil	

11. Spectroscopic studies of N-phenylsulfonyl-a-aminohydrazides

The a-aminohydrazides were identified by NMR spectroscopy at[1] H and C.[13]

11.1NMR spectroscopic properties H[1]

NMR spectroscopy at[1] H has enabled us to confirm the structures of the a-aminohydrazides prepared (see the experimental section). Let us study the case of *N*- phenylsulfonyl- 2-aminopropanehydrazide **90a** as an example.

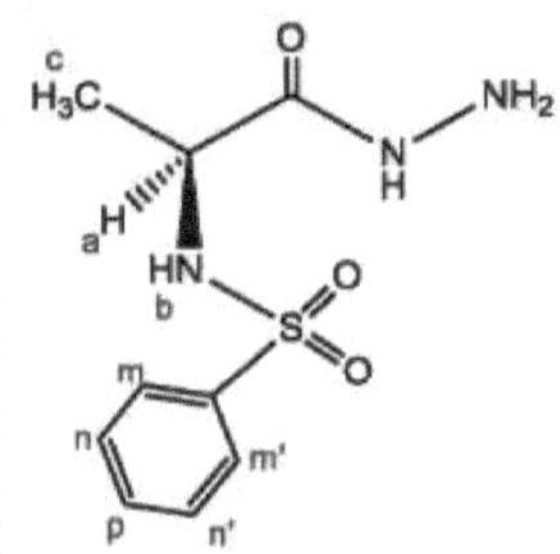

90a

The 1H NMR spectrum of composë **90a** (**Figure 16**)recorded at 300 MHz in deutërëprë chloroform shows the following signals:

- A doublet at 1.27 ppm 3H inegration relating to mëthylic protons (CHs)cavecoupling constant (J= 6.9Hz).
- A quadruplet has 3.87 ppm inegration of a proton relative to the Ha proton with coupling constants (J= 6.9 Hz).
- A bulk between 7.51-7.64 ppm of 3H inegration relating to the Hn, Hn' and Hp protons.
- A doublet at 7.89 ppm 2H integration related to the H protons$_m$ and Hm' with a coupling constant (J=7.2 Hz).

11.2NMR spectroscopic properties C[13]

The total proton dëcoupling NMR spectrum[13] C of the same compoundë **90a**(**Figure 17**)recorded in deutërë chloroform at 75 MHz confirms the structure based on the following signals:

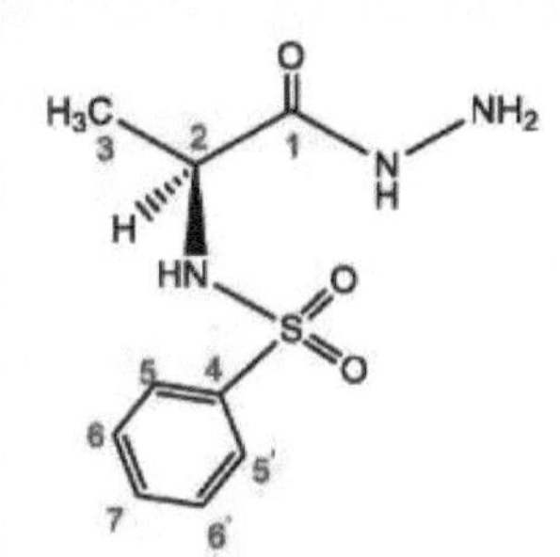

90a

- A signal at 18.76 ppm relating to carbon C_3.
- A signal at 51.70 ppm relating to carbon C_2.
- Signals at 127.08; 129.28; 133.07; 139.65 ppm corresponding respectively to the aromatic carbons $C_{6,6'}$; $C_{5,5'}$; C_7; C_4.
- A signal at 172.05 ppm relating to carbon CO.

> The spectrum clearly shows the disappearance of the characteristic mëthoxy peak.

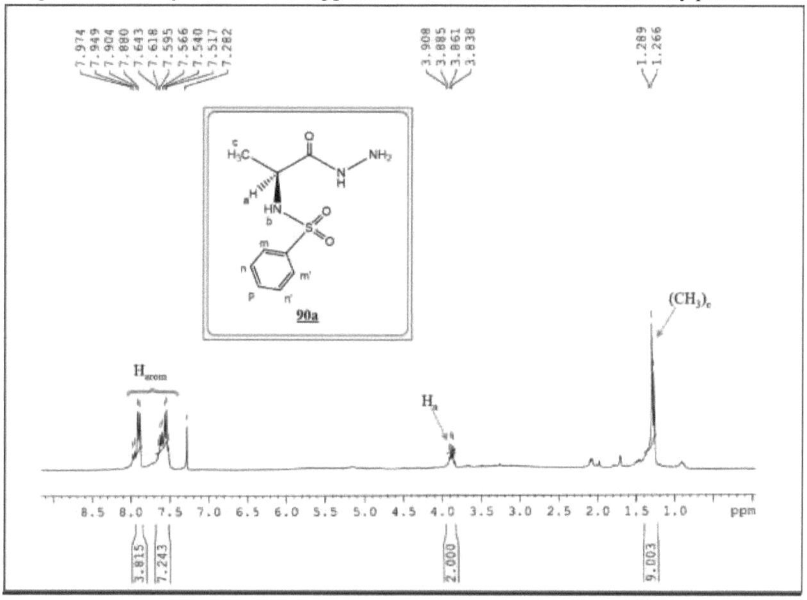

Figure 16. H NMR spectrum of compound 90a recorded at 300 MHz (CDCl 3)

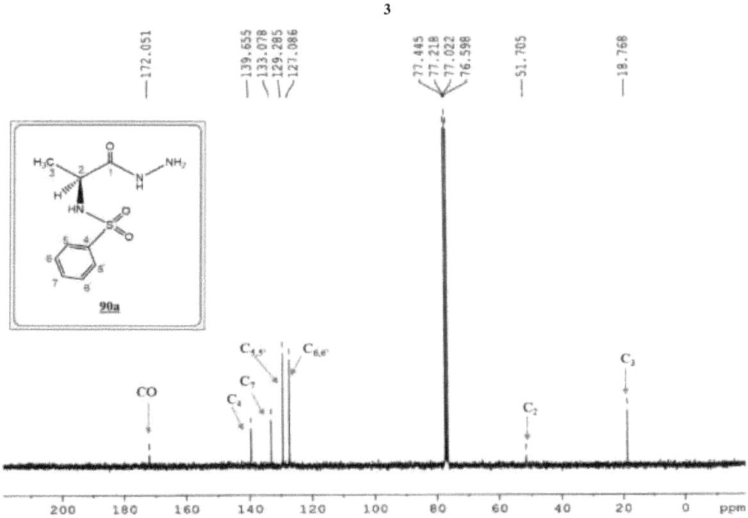

Figure 17. C NMR spectrum of compound 90a recorded in CDCl at 75 MHz

C. Conclusion

In this part of the work we have synthesised a series of six chiral90**(a-f)**a-aminohydrazides in chemical yields of between 76% and 92%. These molecules were ëlë obtained by condensation of the chiral a-aminoesters61**(a-f)**prepared in the second chapter with l'hydrazine in methanol reflux.

General conclusion and outlook

As part of this master's programme, we have described the synthesis of a series of six chiral a-aminohydrazides from natural amino acids. These products are highly sought after thanks to their interesting synthetic properties.

The synthesis strategy adopted in this work consists of first protecting the amine function in order to avoid previous chemioselectivity problems. A series of chiral N-phenylsulphonyl amino acids were obtained in chemical yields ranging from 80% to 93%.

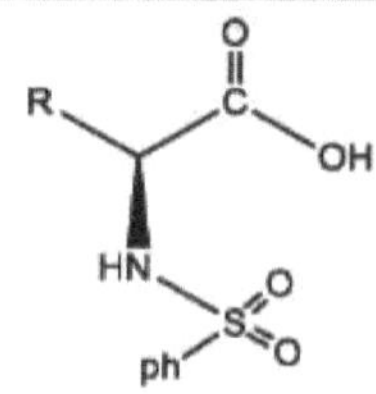

33(a-e)

a : R=Me ; **b** :R= i-Pr ; **c** : R=i-Bu ; **d** : Sec-Bu ; **e** : R=Bn

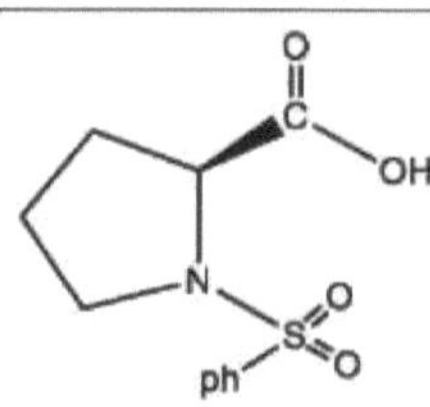

33f

In a second step, we converted the monoproteic amino acids into the corresponding a-amino esters using chlorotrimethylsilane and methanol. The a-aminoesters were obtained in good chemical yields (74% and 95%).

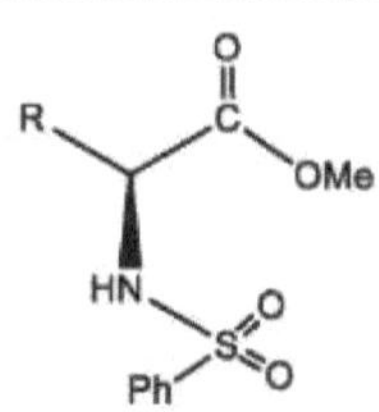

61(a-e)

a : R=Me ; **b** :R= i-Pr ; **c** : R=i-Bu ; **d** : Sec-Bu ; **e** : R=Bn

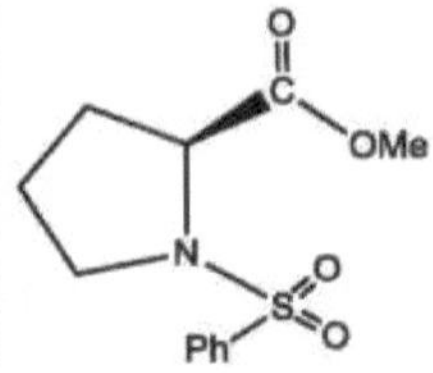

61f

Condensation of the a-aminoesters obtained with l'hydrazine in тё^aпol reflux resulted in the synthetisation ofa sërie of sixchiral a-aminohydrazides in quantitative chemical yields.

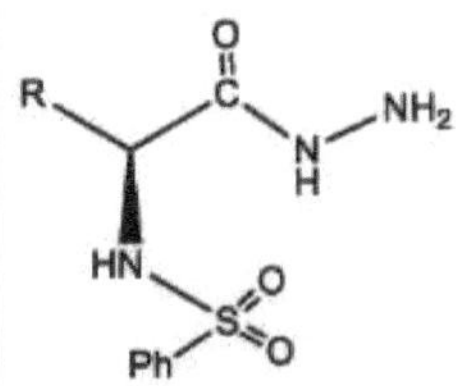

90(a-e)

a : R=Me ; **b** :R= i-Pr ; **c** : R=i-Bu ; **d** : Sec-Bu ; **e** : R=Bn

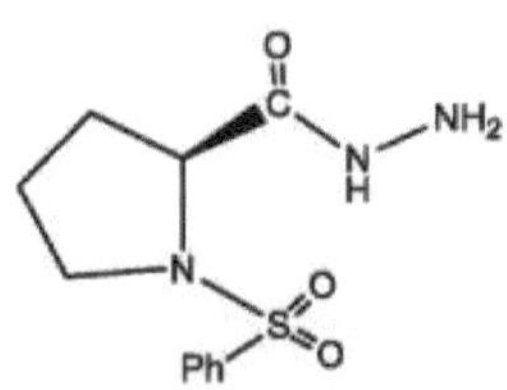

90f

As perspectives of this work, we count:

- Condensing chiral a-aminohydrazides prepared with different bi-electrophiles to produce heterocyclic products of biological interest.
- To determine the Гexcësënantюmërique of our products by reacting it with optically pure acid chlorides.

Experimental part

I. General methods

I.1 Purification of solvents and reagents

The anhydrous solvents are prepared by distillation in an argon atmosphere in the presence of the drying agents.

- Methanol is distilled over magnesium in the presence of a catalytic amount of iodine.

For reagents: thionyl chloride (SOCl2) is distilled under argon just before use.

I.2 Reactions

The reactions, in an anhydrous medium, were carried out under an argon atmosphere and in glassware that was oven-dried just before use. The organic solutions were concentrated under reduced pressure using a rotary evaporator.

I.3 Preparing the developer

The potassium permanganate developer is prepared by dissolving 3g of KMnO4 and 20g of K2CO3 in 300 mL of distilled water in the presence of 5mL of a 5% aqueous sodium hydroxide solution and 2 mL of glacial acetic acid.

I.4 Chromatography

> Thin layer chromatography

Analytical thin-layer chromatography (TLC) was carried out on ready-to-use silica plates (silica gel with Fluka Art 254 nm fluorescent indicator) and revealed using a U.V. lamp (254 nm).

> Column chromatography

Product purification was carried out by column chromatography using silica gel as the stationary phase.

I.5 Analytical methods and equipment

> Nuclear magnetic resonance (NMR)

The NMR spectra were carried out in solution in deuterated chloroform (CDCl3) using a BRUKER AC-300 (300MHz) for ^{1} H NMR and (75 MHz) for ^{13} C NMR.

The internal reference used is tëtramëthylsilane (TMS). The chemical shifts (ppm) are positive towards the low fields compared with TMS and the coupling constants noted J are expressed in Hertz (Hz).

II. Operating procedures

II. 1 Preparation of chiral N-phenylsulfonyl-a-amino acids from natural amino acids

The case of *N-phenylsulfonyl-*(L)-alanine

(L)-alanine (2.8 g, 31.46 mmol, 1 sq.) is dissolved in sodium hydroxide solution (2M, 31.46 mmol), phSnylsulphonyl chloride (4.22 ml, 33.03 mmol, 1.05Sq.) is added at 0°C followed by the addition of diisopropylthylamine (6.03 ml, 34.61mmol, 1.1 ëq.) and acëtone (15.73 ml). After 10 minutes, the ice bath is removed and the mixture is left to stir for 6 hours at room tempërature. Then the тёlапде react ion is washed with disthylic lather. The organic phase is extracted with an aqueous solution of NaOH (2M), then cooled to -10°C and acidified to pH=1 by the addition of concentrated HCl. This aqueous phase is extracted three times with ethyl acetate. The organic phase was washed with a saturated solution of NaCl and sSchSe on MgSO4. After evaporation of the solvent, the residue was recrystallised in a mixture of ethyl acetate and petroleum ether of varying proportions. Compound 33a is obtained.

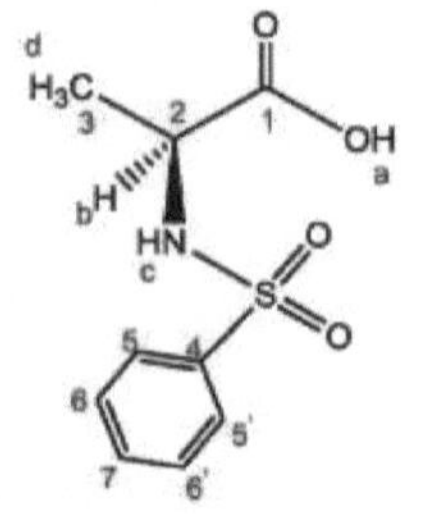

33a

Gross formula: C9H11O4NS
M (g/mol) :229 g/mol
Revealer: UV lamp.
Rdt= 93%
Appearance: white solid
1H NMR (300 MHz, CDCl3); δ (ppm): 1.42 (d, J=7.2Hz,3Hd);4.06 (Q,Ji =14.1 Hz,J2 =6.9 Hz,1Hb) ; 5.44 (d, J = 7.5 Hz, 1Hc) ; 7.28-7.96 (m,5Harom.).
NMR[13] C (75 MHz, CDCl3); δ (ppm): 19.45(C3);51.21(C2);127.07(C6, C6'); 129.10(C5,C5'); 132.90 (C7); 140.00(C4); 176.10 (CO).

The case of *N*- phenylsulfonyl-(L)-valine

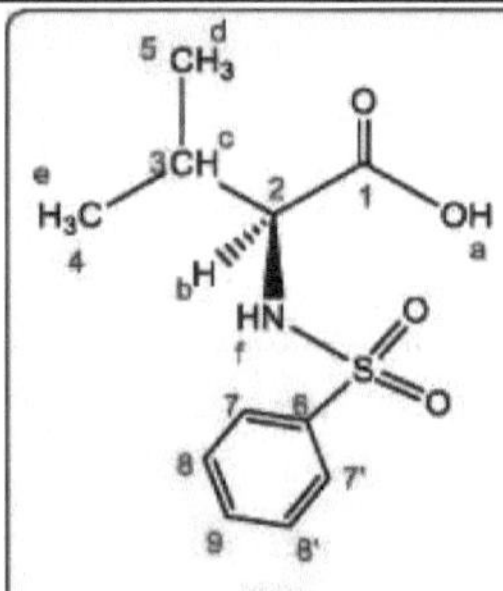

33b

Gross formula: C11H15O4NS
M (g/mol) :257 g/mol
Revealer: UV lamp.
Rdt= 80%
Appearance: white solid
NMR[1] H (300 MHz, CDCl3); δ (ppm): 0.88(d, J=6.9Hz,3He or d); 0.96 (d,J=6.6 Hz, 3Hd or e); 2.04-2.16 (m, 1Hc); 3.81 (dd, J1=9.9 Hz, J2=4.8 Hz, 1Hb);5.15 (d, J =9.9 Hz,1Hf);7.45-7.94(m,5Harom .).
NMR[13] C (75 MHz, CDCl3); δ (ppm):16.69(C4);18.36(C5);30.94(C3); 60.18 (C2);126.73(C8, C8'); 128.49(C7,C7'); 132.38 (C9); 139.32(C6); 174.70 (CO).

The case of N-phenylsulfonyl-(L)-leucine

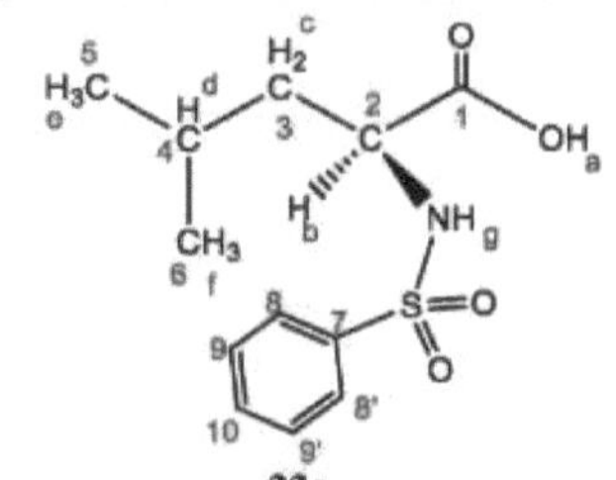

33c

Gross formula: C12H17O4NS
M (g/mol) :271 g/mol
Revealer: UV lamp.
Rdt=85%
Appearance: white solid.
1H **NMR (300 MHz, CDCl3); δ (ppm):** 0.82(d, J=6.3 Hz,3He); 0.89 (d,J=6.6 Hz, 3Hf); 1.491.59 (m, 2Hc); 1.69-1.78 (m, 1Hd);3.69 (t,J =6.3 Hz,1Hb);5.31(d, J=8.7Hz,1Hg); 7.46-7.93(m, 5Harom.).
NMR[13] C (75 MHz, CDCl3); δ (ppm): 20.81(C5);22.09(C6);23.86(C4); 41.62 (C3); 53.67(C2); 126.69(C9, C9'); 128.51(C8,C8'); 132.41 (C10); 139.36(C7); 176.01 (CO).

The case of *N-phenylsulfonyl*-(L)-isoleucine

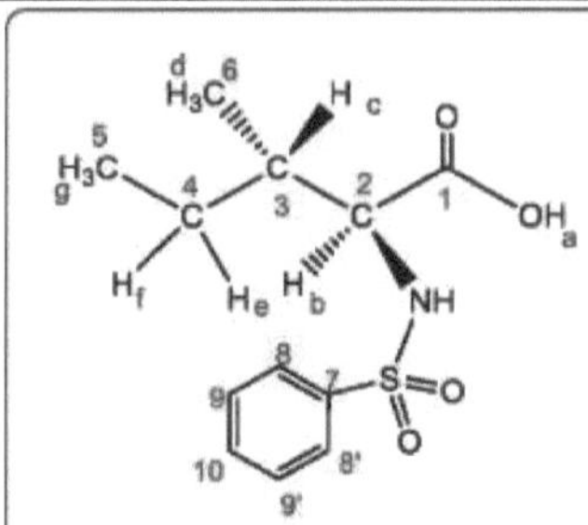

33d

Gross formula: C12H17O4NS
M (g/mol) :271 g/mol
Revealer: UV lamp.
Rdt= 83
Appearance:white solid
1H **NMR (300 MHz, CDCl3); δ (ppm):**0.89(t,J=7.2 Hz,3Hg); 0.93 (d,J=6.9 Hz, 3Hd); 1.131.29 (m, 1Hf or e); 1.39-1.48 (m, 1He or f);1.79-1.87(m,1Hc); 3.88 (s, 1Hb); 7.47-7.60 (m, 5Harom.);8.23 (sl, NH).
NMR[13] C (75 MHz, CDCl3); δ (ppm):11.16(C5);15.30(C6);24.58(C4); 38.24 (C3); 59.97(C2); 127.20(C9, C9'); 128.98(C8,C8'); 132.86 (C10); 139.84(C7); 175.66 (CO).

The case of *N*- phenylsulfonyl-(L)-phenylalanme

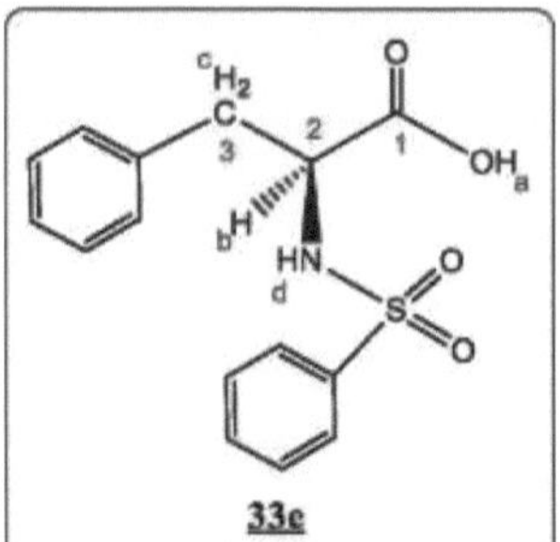

33e

Gross formula: C15H15O4NS
M (g/mol) :305 g/mol
Revealer: UV lamp.
Rdt=91%.
Appearance:white solid
NMR[1] H (300 MHz, CDCl3); δ (ppm): 2.94-3.15(m, 2Hc); 4.27(s,1Hb); 5.16 (sl, 1Hd); 7.09 7.91 (m,10Harom.).
NMR[13] C (75 MHz, CDCl3); δ (ppm): 38.88(C3); 56.44(C2);127.02-139.71(12Carom.); 174.78 (CO).

The case of *N-phenylsulfonyl*-(L)-proline

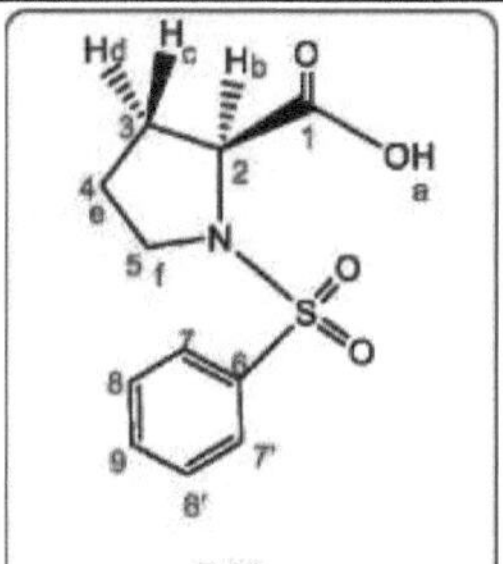

33f

Gross formula: C11H13O4NS.
M (g/mol) :255 g/mol.
Revealer: UV lamp.
Rdt= 80%.
Appearance: white solid.
1H NMR (300 MHz, CDCl3); δ (ppm): 1.55 (m, 1Hd); 1.81 (m, 2He and 1H); 3.34(m, 2Hf);4.26(s,1Hb); 7.55-7.91 (m,5Harom.).
NMR[11 12 13] C (75 MHz, CDCl3); δ (ppm): 31.03(C4);32.30(C3); 47.93(C5); 49.99(C2); 127.95(C7, C7'); 128.35(C8,C8'); 134.61 (C9); 142.86(C6); 177.33 (CO).

1 .2 Preparation of N-phenylsulfonyl-a-aminoesters

The case of methyl *N-phenylsulfonyl-2-aminopropanoate*

Under an inert atmosphere of argon, 6g of *N*- phenylsulfonyl-(L)-alanine (26.2 mmol) is introduced into a 100 ml rodë monocole flask. TMSCl (6.65ml, 52.4mmol, 2 ëq.) is added then mëthanol is added. The mëlange is left to stir for 72 hours at room temperature. After evaporation of the mëthanol and cooling, compound **61a** is obtained.

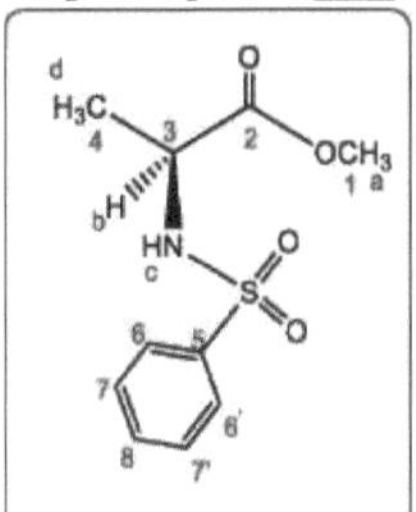

61a

Gross formula: C10H13O4NS

M (g/mol) :243 g/mol

Revealer: UV lamp.

Rdt= 94%

Appearance: white solid

1H **NMR (300 MHz, CDCl3); 8 (ppm):** 1.38(d, J=7.2 Hz,3Hd);3.54 (s,3Ha);4.02 (Q,J1=14.7 Hz,J2 =7.2 Hz,1Hb); 5.24 (d, J = 7.2 Hz, 1Hc); 7.47-7.86 (m,5Harom.).

NMR[13] C (75 MHz, CDCl3); 8 (ppm): 19.24(C4);51.01(C3); 51.96(C1); 126.65(C7, C7'); 128.50(C6,C6'); 132.21 (C8); 139.58(C5); 171.95 (CO).

The case of methyl *N*- phenylsulfonyl-2-ammo-3-methylbutanoate

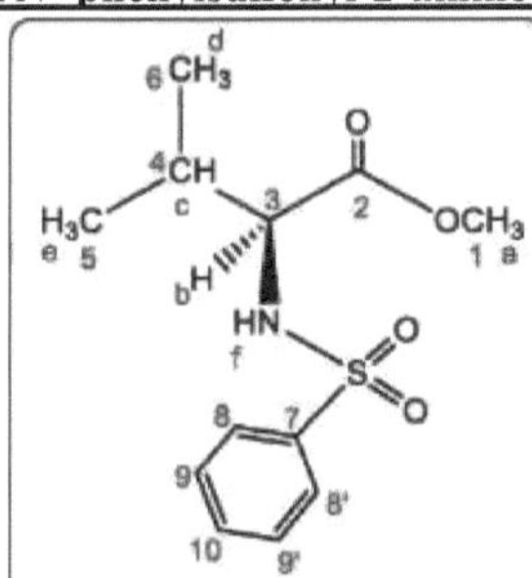

61b

Gross formula: C12H17O4NS

M (g/mol) :271 g/mol

Revealer: UV lamp.

Rdt=74

Appearance: white solid

1H **NMR (300 MHz, CDCl3); 8 (ppm):** 0.88(d, J=6.6 Hz,3He); 0.94 (d,J=6.9 Hz, 3Hd); 1.992.05 (m, 1Hc); 3.44 (s, 3Ha); 3.75 (dd, J1=9.9 Hz,J2=5.1 Hz, 1Hb);5.10 (d, J =9.9 Hz,1Hf);7.45-7.84(m,5Harom.).

NMR[13] C (75 MHz, CDCl3); 8 (ppm): 16.96(C5);18.28(C6);31.15(C4); 51.48 (C1); 60.66 (C3);

126.77(C9, C9'); 128.38(C8,C8'); 132.12 (C10); 139.42(C7); 171.08 (CO).

The case of methyl N-phenylsulfonyl-2-amino-4-methylpentanoate

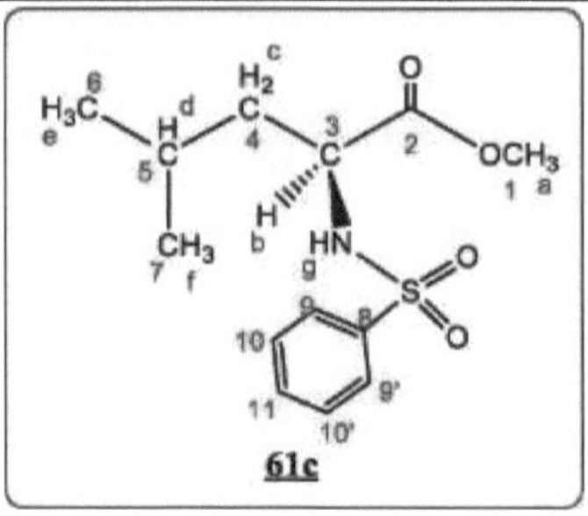

Gross formula: C13H19O4NS

M (g/mol) :285 g/mol

Revealer: UV lamp.

Rdt= 95%

Appearance:white solid

NMR[1] H (300 MHz, CDCl3); δ (ppm): 0.83(d, J=6.3 Hz,3He); 0.91 (d,J=6.6 Hz, 3Hf); 1.491.53 (t, J=7.2Hz, 2Hc); 1.72-1.83 (m, 1Hd);3.46(s, 3Ha);3.97 (s,1Hb); 7.48-7.87(m, 5Harom.).

NMR[13] C (75 MHz, CDCl3); δ (ppm): 20.94(C6);22.09(C7);23.83(C5); 41.94 (C4); 51.65(C1); 53.97 (C3); 126.78(C10, C10'); 128.42(C9,C9'); 132.17 (C11); 139.36(C8); 172.03 (CO).

The case of methyl N-phenylsulfonyl-2-amino-3-methylpentanoate

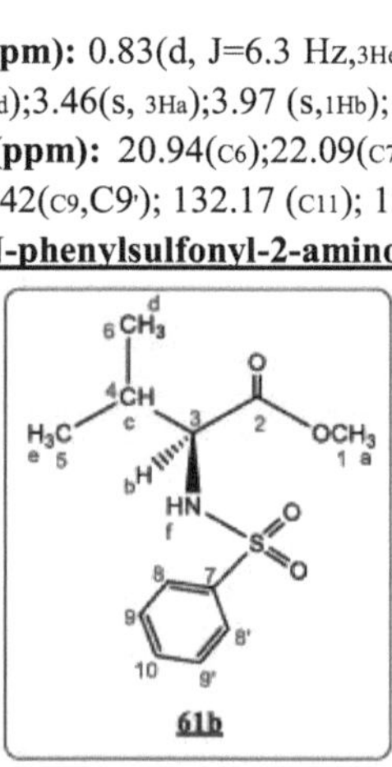

Gross formula: C13H19O4NS

M (g/mol) :285 g/mol

Revealer: UV lamp.

Rdt= 88%

Appearance: white solid

1H NMR (300 MHz, CDCl3); δ (ppm): 0.84-0.90 (m, 6H, (CH3)d and (CHs)g); 1.10-1.20 (m, 1Hf or e); 1.38-1.46 (m, 1He or f); 1.70-1.80 (m,1Hc); 2.16 (s, 3Ha); 3.79 (dd, J=9.9Hz,J=5.7 Hz, 1Hb); 5.14 (d, J=9.9Hz,1Hh); 7.46-7.83 (m, 5Harom.).

NMR[13] C (75 MHz, CDCl3); δ (ppm): 10.69 (C6);14.82 (C7);24.15 (C5); 37.90 (C4); 51.60 (C1); 59.77 (C3); 126.79 (C10, C10'); 128.47 (C9,C9'); 132.27 (C11); 139.11 (C8); 171.17 (CO).

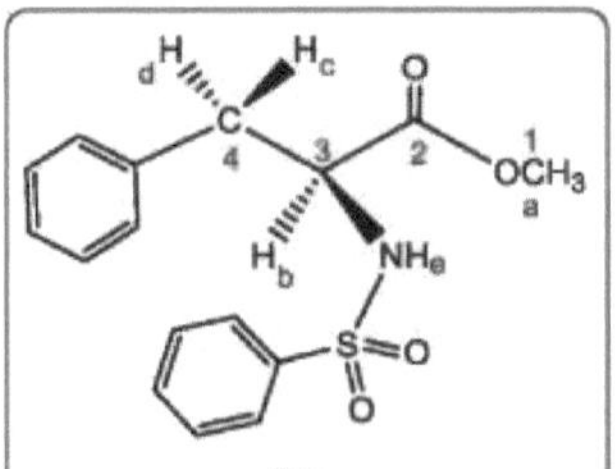

61e

Gross formula: C16H17O4NS
M (g/mol): 319 g/mol
Revealer: UV lamp.
Rdt=95
Appearance: white solid
1H **NMR (300 MHz, CDCl3); 8 (ppm):** 3.04(d, J=6 Hz, 2Hc,d); 3.49 (s, 3Ha); 4.24 (q,J1=14.1Hz,J2=6 Hz,1Hb); 5.04(d, J=8.4Hz, 1He); 7.06-7.77 (m,10Harom.).
NMR[13] C (75 MHz, CDCl3); 8 (ppm): 38.96(C4);51.75(OCH3);56.24(C3);126.63-139.44(12Carom.); 170.62 (CO).

<u>The case of methyl *N-phenylsulfonylpyrrolidine-2-carboxylate*</u>

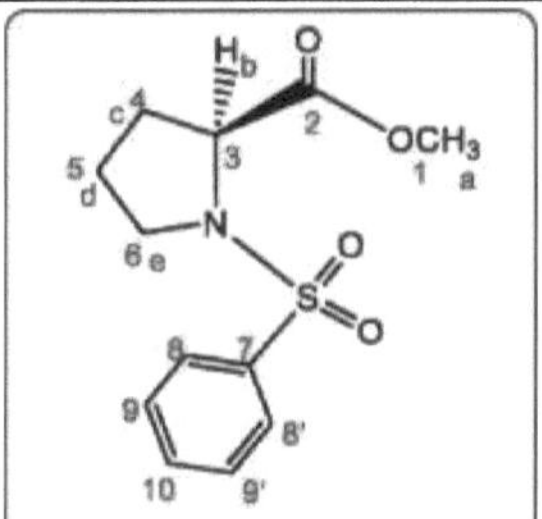

61f

Crude formula: C12H15O4NS.
M (g/mol) :269 g/mol.
Revealer: UV lamp.
Rdt= 95%.
Appearance: white solid.
1H **NMR (300 MHz, CDCl3); 8 (ppm):** 1.81(m, 2Hd); 2.02 (m, 2Hc); 3.45(m,2He); 3.72(s, 3Ha);4.35(s,1Hb); 7.55-7.91 (m,5Harom.).
NMR[13] C (75 MHz, CDCl3); 8 (ppm):24.21(C5);30.47(C4); 48.02(C6); 52.05 (C1); 59.92(C3); 126.98(C8, C8'); 128.58(C9,C9'); 132.33 (C10); 137.78(C7); 172.05 (CO).

II .3 Preparation of N-phenylsulfonyl-a-aminohydrazides

<u>The case of N-phenylsulfonyl-2-amino-4-methylpentanehydrazide</u>

To 2.45g of **<u>mëthyl61c-N-phenylsulphonyl-2-amino-4-</u>** mëthylpentanoate (8.59mmol) and 1'hydrazine (5 ëq, 42.95mmol) in a refluxing methanol flask is added.

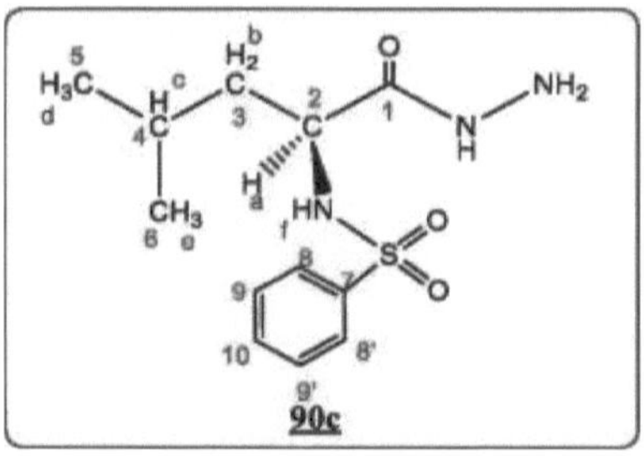

Gross formula: C12H19O3N3S
M (g/mol) :2b85 g/mol
Revealer: UV lamp.
Rdt= 92
Appearance: white solid
NMR[1] H (300 MHz, CDCl3); 8 (ppm): 0.65(d, J=5.1 Hz,3Hd); 0.80 (d,J=5.4 Hz, 3He); 1.421.54 (m, 1Hc, 2Hb);3.76 (s,1Ha);5.90(sl,1Hf); 7.48-7.89(m, 5Harom.).
NMR[13] C (75 MHz, CDCl3); 8 (ppm): 21.28(C5);22.67(C6);24.30(C4); 42.04 (C3); 54.54 (C2); 127.24(C9, C9'); 129.09(C8,C8'); 132.91 (C10); 139.83(C7); 172.15 (CO).

The case of N-phenylsulfonyl-2-aminopropanehydrazide

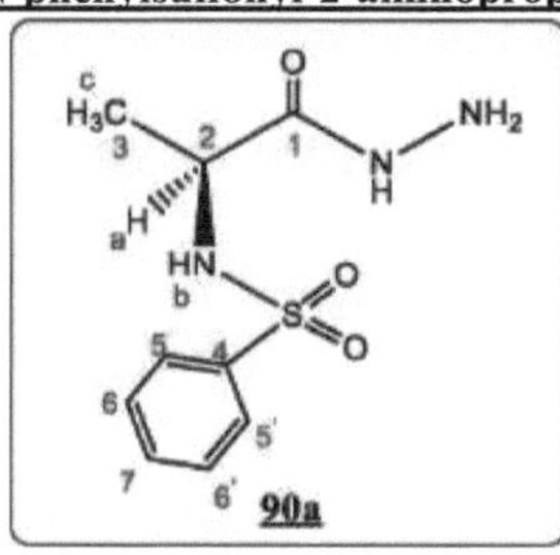

Gross formula: C9H13O3N3S
M (g/mol):243 g/mol
Developer: UV lamp.
Rdt=76%.
Appearance: white solid
1H NMR (300 MHz, CDCl3); 8 (ppm): 1.27(d, J=6.9 Hz,3Hc); 3.87 (q,Ji=14.1Hz, J2=6.9Hz,1Ha); 7.51-7.97 (m,5Harom.).
NMR[13] C (75 MHz, CDCl3); 8 (ppm): 18.76(C3);51.70(C2); 127.08(C6, C6'); 129.28(C5,C5'); 133.07 (C7); 139.65(C4); 172.05 (CO).

The case of *N*- phenylsulfonyl-2-amino-3-methylbutanehydrazide

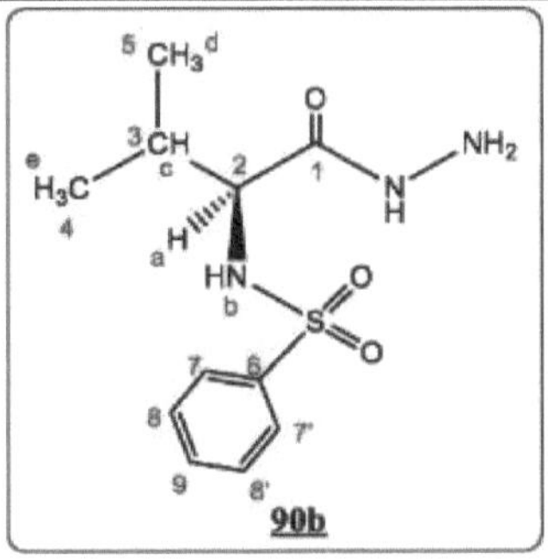

Gross formula: C11H17O3N3S
M (g/mol) :271 g/mol
Revealer: UV lamp.
Rdt=87%.
Appearance: white solid
1H **NMR (300 MHz, CDCl3); 8 (ppm):** 0.79(d, J=6.9 Hz,3Hor e); 0.85 (d,J=6.9 Hz, 3Heor d); 1.75-2.05 (m, 1Hc); 3.48 (d, J=6 Hz, 1Ha);7.46-7.85(m,5Harom.).
NMR[13] C (75 MHz, CDCl3); 8 (ppm): 17.21(C4);18.36(C5);30.71(C3); 60.91(C2);126.69(C8, C8'); 128.48(C7,C7'); 132.31 (C9); 139.43(C6); 170.72 (CO).

The case of N-phenylsulfonyl-2-ammo-3-methylpentanehydrazide

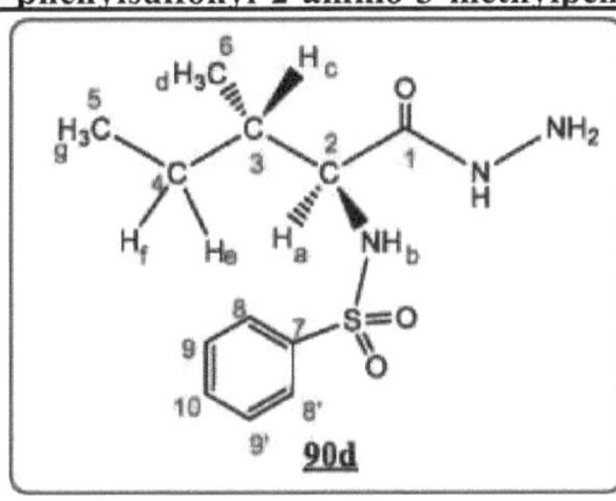

Gross formula: C12H19O3N3S
M (g/mol) :285 g/mol
Revealer: UV lamp.
Rdt=81%
Appearance: white solid
1H **NMR (300 MHz, CDCl3); 8 (ppm):** 0.78(t, J=6.9 Hz,3Hg); 0.84 (d,J=7.2Hz, 3Hd); 1.001.12 (m, 1Hf or e); 1.41-1.54 (m, 1He or f);1.71-1.77(m,1Hc); 3.51 (d, J=6Hz, 1Ha); 7.48-7.85 (m, 5Harom.).
NMR[13] C (75 MHz, CDCl3); 8 (ppm): 11.02(C5); 15.23 (C6);24.58 (C4); 37.79 (C3); 60.46(C2); 127.26(C9, C9'); 129.03(C8,C8'); 132.89 (C10); 139.77(C7); 171.03 (CO).

The case of *N*- phenylsulfonyl-2-ammo-3-phenylpropanehydrazide

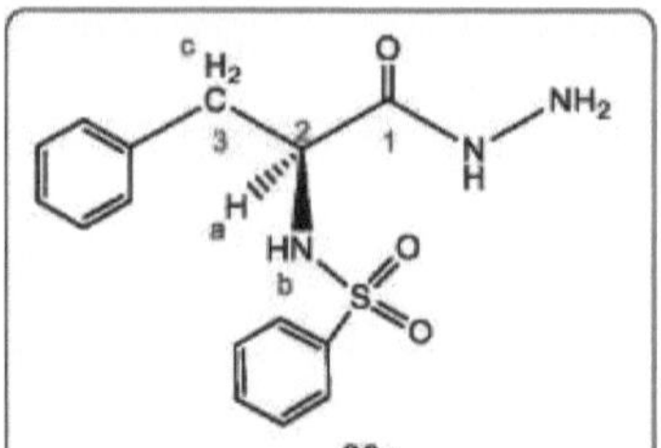

90e

Gross formula: $C_{15}H_{17}O_3N_3S$
M (g/mol) :319 g/mol
Revealer: UV lamp.
Rdt=78
Appearance: white solid
NMR[1] H (300 MHz, CDCl3); 8 (ppm): 2.96(s, 2Hc); 4.60-4.82 (q,J1 =14.1 Hz, J2=6 Hz,1Ha); 5.59(d, J=12 Hz, 1Hb); 7.13-7.64 (m,10Harom.).
NMR[13] C (75 MHz, CDCl3); 8 (ppm): 38.52(C3); 56.24(C2); 127.03-139.21(12Carom.); 170.62 (CO).

The case of N-phenylsulfonyl pyrrolidine-2-carbohydrazide

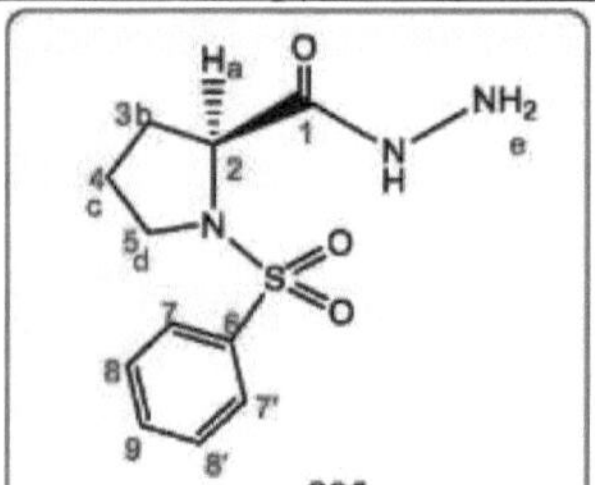

90f

Gross formula: $C_{11}H_{15}O_3N_3S$
M (g/mol) :269 g/mol
Developer: UV lamp
Rdt= 85%.
Appearance:yellowish viscous oil
1H NMR (300 MHz, CDCl3); 8 (ppm): 1.22-1.55(m, 2Hc); 1.76-2.05 (m, 2Hb); 3.06-3.52(m, 2Hd); 4.12-4.15(m, 1Ha); 4.65 (s, 2He); 7.42-7.81 (m, 5Harom.).
NMR[13] C (75 MHz, CDCl3); 8 (ppm): 23.81(C4);29.88 (C3); 49.11(C5); 61.10 (C2); 127.21 (C7, C7'); 128.87 (C8,C8'); 132.88 (C9); 135.32 (C6); 171.28 (CO).

References

1- M. Marinozzi, Bioorg. Med. Chem, **2013,** 21, 3780.
2- J. L. Bada, S. L. Miller,BioSystems,**1987**, 20, 21.
3- J. Clayden, N. Greeves, N. Warren, S. P. Wothers, **2002,** 45, 1219.
4- H. Kagan, "Catalysis. Asymetrique, Pour la science, Paris, **1992**, 172, 42.
5- M. Lahlou, Flavour Fragrance J, **2004**, 19, 159.
6- J. C. R. Gongalves,F. Oliveira,R. B. Benedito, D. P.R. Almeida,D. A. M. Araujo, Biol. Pharm. Bull, **2008,**31, 1017.
7- F.V.M. Souza, M. B. Rocha, D. P. de Souza, R. M. Margal, Fitoterapia, **2013**, 85, 20.
8- G. Buchbauer, W. Jager, A. Gruber, H. Dietrich, FlavourFragr. J., **2005**, 20, 686.
9- L. Chiodini, G. Pepeu, Composition comprising S-oxiracetam for use as nootropic: WO, 9306826 (P).
10- Zhang, Q. Zhang, W. Yang, Q. Zhang, Y. Yang, J. Li, Y. Lu, Y. Zheng, J. He, D. Zhao, X. Chen, J. Chromatogr. **2015**, B 993-994, 9.
11- R. N. Pechnick, R. E. Poland, JPET, **2004**, 309, 515.
12- Y. Qi, D. Liu, W. Zhao, C. Liu, Z. Zhou, P. Wang, Pesticide Biochemistry and Physiology, **2015**, 01-07. In press (http://dx.doi.org/10.1016/j.pestbp.2015.06.004).
13- G. M. Coppola, H. F. Schuster, Assymetric synthesis: " Construction of chiral molecules using amino acids", Weily ; New York, **1987.**
14- G. C. Barrett, Chemistry and Biochemistry of the Amino Acids, Chapman and Hall, London, **1985**.
15- J. Jackli, C. Limberakis, Modern organic synthesis in the laboratory, *Oxford univ, USA,* **2007**.
16- R. M. Williams, "Synthesis of optically active a-aminoacids"; pergamon:Oxford, **1989**.
17- M. Mikalajczyk, P. Balczewski, *Synthesis, 659*, **1987**.
18- J. Domagala, J. Wemple, *Tetrahedron Lett,***1973**, *14*, 1179.
19- F. A. Davis, B. Yang, J. Deng, J. *Org Chem,* **2003**, *68*, 5147.
20- L.A. Caprino, J. Am, *Chem, Soc,* **1957**, *79*,98.
21- F. C. Mckay, N. F. Albertson, J. Am, *Chem Soc, **1957**, 79*, 6186.
22- S. Tchertchian, O. Hartley, P. Botti, J. *Org Chem,* **2004**, *69*, 9208.
23- E. At herton, R. C. Sheppar, *Academic Press: New York,* **1987**, *9*, 1.
24- M. Bergmann, L. Zervas, *Ber,* **1932**, *65,*1192.
25- R. H. Sifferd, Duvigneaud, *J.Biol. Chem,* **1935,** *108,* 753.
26- D. A. Kidd, F. E. King, *Nature,* **1948**,62, 776.
27- A. Rougny, M. Daudon, *Chem Soc,* **1976**, *5*, 833.
28- B. B. Malcoln, C. Donald craig, Synlett.,**1992**, 41.
29- P. Strazzolini, P. Strazzolini, T. Melloni, A. G. Giumanini, Tetrahedron, **2001**, 57, 9033.
30- N. J. Tom, W. M. Simon, H. N. Frost, M. Ewing, Tetrahedron Lett,**2004**, 45, 905.
31- S. El Kazzouli, J. Koubachi, S. B. Raboin, A. Mouaddibb, G. Guillaum, Tetrahedron Lett,**2006**, 47, 8575.
32- P. R. Sultane, P. R. Sultane, T. B. Mete, R. G. Bhat, Tetrahedron Lett, **2015,** 56, 2067.
33- H. Hidemasa, Y. Yokoyama, Synthesis, **2000**, 21.
34- T. L. Cupps, R. H. Boutin, H. Rapoport, J. Org Chem, **1985**, 50, 3972.
35- T. F. Buckley, H. Rapoport, J. Am. Chem. soc. **1981**, 103, 6157.
36- Olah, G. A. In "Friedel-Crafts and Related Reactions"; Interscience: New York, **1964**;

Vols. I-IV.
37- Z. G. Hajos, D. R. Parrish, J. Org. Chem. **1974**, 39.
38- G. L. Buchanan,Chem.Soc. Rev.,**1988**, 17**,** 91.
39- H. D. Dakin, R. West, J. Biol. Chem, **1928**, 78**,** 745.
40- N. Tka, J. Kraiem, Y. Kacem, A. Hajri, B. B. Hassine, C.R.Chimie, **2009,** 12 1066.
41- L. W. Bieber, M. C. F. de Araujo,Molecules, **2002**, 7, 902.
42- R. M. Freidinger, J. S. Hinkle, J. Org Chem, **1983**, 48, 77.
43- G. C. Barrett, 'Amino Acids, Peptides and Proteins' R.S.C. Publications: London **1996**, *27*, 1.
44- S. L. Manjinder, , K.R. Yeeman, N. G. J. Michael, C. V. John, Serine, *J. Org. Chem,* **2002**, *67,* 1536.
45- V. K. Tandon, D. B. Yadav, R. V. Singh, A.K. Chaturvedi, , P. K. Shukla, *Bioorg. Med. Chem.Lett,.***2005**, *15*, 5324.
46- A. Q. Hussein, M. M. El-Abadelah, W. S. Sabri, J. Heterocyclic. Chem, **1984**, 21, 455.
47- C. Lamberth, Tetrahedron, **2010,** 66, 7239.
48- U Balakrishnan, N Ananthiet S Velmathi, Indian. J. Chem, **2011**, 50, 1157.
49- . J. Li, Y. Sha, Molecules, **2008,** 13, 1111.
50- A. Kudelko, W. Zielin'ski, Tetrahedron, **2009**, 65, 1200.
51- K. A. Kumar, P. Jayaroopa, G. V. Kumar, Int. J. ChemTech Res, 2012, 4, 1782**.**
52- K. Du, X. Cao, P. Zhang, H. Zheng, Bioorg. Med. Chem. Lett, **2014**, 24, 5318.
53- H. Y. Ma, Chin. Chem. Lett, **2013,** 24, 780.
54- A. K. Yadav, L. D. S. Yadav, Tetrahedron Lett.,**2014,** 55, 2065.
55- I. N. Arthur, J. E. Hennessy, D. Padmakshan, D. J. Stigers, S. Lesturgez, S. A. Fraser, M. Liutkus, G. Otting, J. G. Oakeshott, C. J. Easton, J. European Chem. **2013**, 19, 6824.
56- H. A. Abdel-Aziz, N. A. Hamdy, A. M. Farag, I. M. I. Fakhr, J. Chinese Chem. Soc. **2007**, 54, 1573.
57- S. A. Shahzad, M. Yar, M. Bajda, B. Jadoon, Z. A. Khan, S. A. Naqvi, A. J. Shaikh, K. Hayat, A. Mahmmod, N. Mahmood, S. Filipek, Bioorg. Med. Chem, **2014**, 1008, 22.
58- S. Zhang, Y. Luo, L. Q. He, Z. J. Liu, A. Q. Jiang, Y. H. Yang, H. L. Zhu, Bioorg.Med. Chem, **2013**, 21, 3723.
59- F. Zhang, X. L. Wang, J. Shi, S. F. Wang, Y. Yin, Y. S. Yang, W. M Zhang, H.L. Zhu, Bioorg. Med. Chem, **2014**, 22, 468.
60- P. Li, L. Shi, X. Yang, L. Yang, X. W. Chen, F. Wu, Q. C. Shi, W. M. Xu, M. He, D. Y. Hu, B. A. Song, Bioorg. Med. Chem. Lett, **2014**, 24, 1677.
61- N. N. Farshori, M. R. Banday, A. Ahmad, A. U. Khan, A. Rauf, Bioorg. Med.Chem. Lett, **1933**, 2010, 20.
62- N. C. Desai, N. Bhatt, H. Somani, A. Trivedi, Eur. J. Med. Chem, **2013**, 67, 54.
63- R. A. Rane, P. Bangalore, S. D. Borhade, P. K. Khandare, Eur. J. Med. Chem, **2013**, 70, 49.
64- H. Kandemir, C. Ma, S. K. Kutty, D. S. Black, R. GriffithP. J. Lewis, N. Kumar, Bioorg. Med. Chem, **2014**, 22, 1672.
65- I. Seidel, W. Plass, D. Klemm, J. Becher, Tetrahedron, **2006,** 62, 5675.
66- A. Manvar, A. Bavishi, A. Radadiya, J. Patel, V. Vora, N. Dodia, K. Rawal, A. Shah,Bioorg. Med. Chem. Lett, **2011,** 21, 4728.
67- Q. F. Liang, Q. F. Liang, J. Liu, J. Chen, TetrahedronLett, **2011**, 52, 3987.
68- D. Perdicchia, E. Licandro, S. Maiorana, C. Baldoli C. Giannini, Tetrahedron,**2003,**59,

7733.
69- A. Kudelko, Tetrahedron, **2012**, 68, 3616.
70- N.I. Arthur, J. E. Hennessy, D. Padmakshan, D. J. Stigers, S. Lesturgez, S. A. Fraser, M. Liutkus, G. Otting, J. G. Oakeshott, C. J. Easton, Chem. Eur. J. **2013**, 19, 6824.
71-

MIX
Papier aus verantwortungsvollen Quellen
Paper from responsible sources
FSC® C105338

Printed by Books on Demand GmbH, Norderstedt / Germany